Lecture Notes in Mathematics

Edited by A. Dold and B. Eckmann

Subseries: Harvard/MIT

Adviser: G. Sacks

574

Johan Moldestad

Computations in Higher Types

Springer-Verlag

Berlin · Heidelberg · New York 1977

Author
Johan Moldestad
Matematisk Institutt
Postboks 1053
Blindern, Oslo 3/Norway

Library of Congress Cataloging in Publication Data

Moldestad, Johan, 1946-
Computations in higher types.

(Lecture notes in mathematics ; 574)
Bibliography: p.
Includes index.
1. Recursive functions. I. Title. II. Series: Lecture notes in mathematics (Berlin) ; 574.
QA3.L28 no. 574 [QA248.5] 510'.8s [511'.3]
77-1375

AMS Subject Classifications (1970): 02F27, 02F29

ISBN 3-540-08132-1 Springer-Verlag Berlin · Heidelberg · New York
ISBN 0-387-08132-1 Springer-Verlag New York · Heidelberg · Berlin

Printed in Germany

Printing and binding: Beltz, Offsetdruck, Hemsbach/Bergstr. 2141/3140-543210

Preface

§§ 1-9 and § 13 have previously appeared in the Preprint Series of the Institute of Mathematics, University of Oslo (No. 2 1976 Computation Theories on two Types by Johan Moldestad). A few corrections and changes have been made. §§ 10-11 is a joint work of Dag Normann and myself. In § 10 the notion of a countable recursion structure is due to him, and independently to Jan Bergstra [23]. Part A in § 11 is due to both of us. Part B is written by me, and part C by Dag Normann. § 12 is written by me in collaboration with professor Jens Erik Fenstad.

Jens Erik Fenstad inspired me to write this book. He has given valuable comments concerning selection of topics, improvements, corrections, etc. I owe him great debt for this and for the interest he has shown in the work. I am also grateful to Dag Normann for helpful discussions, and to Randi Møller and Sigrid Cordtzen for the typewriting.

Oslo, 1976

Johan Moldestad

Contents

Abstract

Recursion in higher types was first studied by Kleene. In his fundamental paper "Recursive functionals and quantifiers of finite type I, II" [12] he introduced nine schemata which generate the partial recursive functions of higher types. The domain of a partial recursive function is a subset of a cartesian product $Tp(i_1) \times \ldots \times Tp(i_n)$ where $i_1 \ldots i_n \in \omega$. $Tp(i)$ is defined by: $Tp(0) = \omega$, $Tp(n+1)$ is the set of functions from $Tp(n)$ to ω. A great deal of effort has been devoted to the study of these functions, and functions closely related to them, in particular the set $\{\varphi :$ For some φ' which is partial recursive $\varphi = \lambda a\, \varphi'(a,F)\}$, where a is a variable ranging over some cartesian product of types $\leq n$, and F is a fixed object to type $n+2$.

The case $F = {}^2E$ has been of particular interest. The nine schemata for recursion in ${}^2E$ generate a hierarchy for the Δ^1_1 subsets of ω ([12]). The case F is of type 2 and normal (i.e. ${}^2E$ is recursive in F) also generates a nice hierarchy ([22]). Many of the deeper properties of recursion in a normal type 2 functional is due to R. Gandy [4], in particular the prewellordering property of computations and the existence of a selection operator.

There is a basic difference between the two cases $F = {}^2E$ and $F = {}^3E$. The last case was first studied by Moschovakis [14], and it was he that first observed that the relations which are recursively enumerable in ${}^3E$ are not closed under existantial quantifiers ranging over ${}^{\omega}\omega$. They are, however, closed under numerical quantifiers.

The case where F is of type $n+2$, $n > 0$, and F is normal (i.e. ${}^{n+2}E$ is recursive in F), is in many ways similar to the case $F = {}^3E$. After some poineering work by Grilliot in his thesis and subsequent papers [5], [6], the case has recently been extensively studied in the theses of Harrington [7] and MacQueen [13]. In [13] it is proved that the relations which are recursively enumerable in F are not closed under existential quantifiers ranging over Tp(n), but they are closed under existential quantifiers ranging over Tp(n-1). So in this connection Tp(n) plays the same role as the reals do in recursion in ${}^3E$. Tp(n-1) plays the role of the numbers. It is natural to introduce the notions individuals and subindividuals. The individuals are the objects of type n, and the subindividuals are the objects of type n-1.

Recursion theory in a normal type 2 object has successfully been characterized in an "abstract" way, either in form of Spector theories or in form of Spector classes. The purpose of this book is to regard recursion in a normal object ot type > 2 as a recursion theory on a domain consisting of two types, viz. the subindividuals and the individuals. A similar approach has been adopted by Harrington and MacQueen in [10]. In the first part (§§ 1-5) we study recursion in a normal functional over a domain of the form $A = S \cup {}^S\omega$, where S is the set of subindividuals and ${}^S\omega$ is the set of individuals. We shall see that most of the results of recursion in normal objects of type $n+2$, $n > 0$, hold in this setting, and that our results specialize to the classical ones when we take $S = Tp(0) \cup \ldots \cup Tp(n-1)$.

In §§ 6-9 we start out with a computation theory on a domain of two types. Computation theories were first studied by Moschovakis [16] and further developed by Fenstad in [2], [3], and we refer to these papers for a survey of the general theory. The aim of this part is to prove the general plus 1 and plus 2 theorem in this setting of two types. For recursion in higher types these results are due to Sacks [21] and Harrington [7]. We also give a characterization of those computation theories which are equivalent to recursion in a normal functional over the domain.

In § 11 we study the gaps which occur in the hierarchies generated by normal functionals on a domain of two types.

We now present a more detailed outline of the paper.

§ 1 The computation domain

The set of subindividuals is denoted by S. We suppose that there is a tuple $\mathcal{S} = (N,s,M,K,L)$ such that $N \subseteq S$ is a copy of the natural numbers with successor function s. M is an injection $S \times S \to S$ (i.e. a pairing function), and K and L are the inverse functions of M, i.e. $K(M(r,s)) = r$, $L(M(r,s)) = s$. From these functions one can define an injection $\bigcup_{n<\omega} S^n \to S$ and a decoding function such that $(\langle x_1 \ldots x_k \rangle)_i = x_i$.

Let $I = S \cup {}^S\omega$, where ${}^S\omega$ denotes the set of functions from S to ω. An element of S is called a subindividual, an element of I is called an individual. In a simple way the functions $M, K, L, \langle\ \rangle, (\cdot)$ can be extended from S to I.

Let $\mathfrak{A} = (I,S,\mathcal{S})$. This is our computation domain. A functional is a function F from I to ω, where ${}^I S$ denotes the set of functions from I to S.

§ 2 Recursion on $\mathfrak{A}$

Let $\mathcal{L}$ be the list $R_1 \ldots R_k$, $\varphi_1 \ldots \varphi_l$, $F_1 \ldots F_m$ where $R_1 \ldots R_k$ are relations on I, $\varphi_1 \ldots \varphi_l$ are partial functions and $F_1 \ldots F_m$ are functionals. We define the functions which are partial recursive in $\mathcal{L}$ by 17 schemata. They are similar to the 9 schemata of Kleene in [12], and to those introduced by Moschovakis in [16]. We define a monotone inductive operator $\Gamma : \bigcup_{n<\omega} I^n \to \bigcup_{n<\omega} I^n$. If $(e,a,r) \in \Gamma(X)$ then $e \in N$, a is a list of length n of objects from I, $r \in S$. The function $\{e\}^{\mathcal{L}}$ is defined by: $\{e\}^{\mathcal{L}}(a) \simeq r$ iff $(e,a,r) \in \Gamma^{\infty}$. f is partial recursive in $\mathcal{L}$ if $f = \{e\}^{\mathcal{L}}$ for some e. In this case e is an index for f. Some of the clauses in the definition of Γ are given below:

I $\quad (\langle 1,n+1 \rangle, x,a,0) \in \Gamma(X)$ if $x \in N$,

$\quad\; (\langle 1,n+1 \rangle, x,a,1) \in \Gamma(X)$ if $x \notin N$.

By this clause the characteristic function of N is recursive. Clauses II - VIII take care of the characteristic function of S, the functions $f(a) = m$ where $m \in N$, the functions s,M,K,L.

IX $(\langle 9,n+2\rangle, x,y,a,x(y)) \in \Gamma(X)$ if $x \in {}^{S}\omega$, $y \in S$

$(\langle 9,n+2\rangle, x,y,a,0) \in \Gamma(X)$ otherwise.

X If $\exists y[(e,a,y) \in X$ and $(f,y,a,x) \in X]$

then $(\langle 10,n,e,f\rangle, a,x) \in \Gamma(X)$.

By clause IX the function $f(x,y) = x(y)$ if $x \in {}^{S}\omega$, $y \in S$, $f(x,y) = 0$ otherwise, is recursive. By X the partial recursive functions are closed under substitution. The clauses XI and XII are for primitive recursion and permutations of the list of arguments. By clause XIII there is a partial recursive function which enumerates all partial recursive functions.

XIII If $(e,a,x) \in X$ then $(\langle 13,n+1\rangle, e,a,x) \in \Gamma(X)$.

The clauses XIV - XVI are for the list $\mathcal{L}$.

XIV $(\langle 14,j_i+n,i\rangle, b,a,0) \in \Gamma(X)$ if $R_i(b)$

$(\langle 14,j_i+n,i\rangle, b,a,1) \in \Gamma(X)$ if $\neg R_i(b)$

$i = 1 \ldots k$, b has length j_i.

XVI If $\forall x \exists y(e,x,a,y) \in X$ then

$(\langle 16,n,e,i\rangle, a, F_i(f)) \in \Gamma(X)$ where

$f(x) = y$ iff $(e,x,a,y) \in X$, $i = 1 \ldots m$.

XVII If $\forall x \in S$ $\exists y(e,x,a,y) \in X$ and

$(e',z,a,u) \in X$ then $(\langle 17,n,e,e'\rangle,a,u) \in \Gamma(X)$, where

$z \in {}^{S}\omega$ is defined by: $z(x) = y$ iff $(e,x,a,y) \in X$.

The functionals are introduced in XVI. Clause XVII says that if $z \in {}^{S}\omega$ is recursive in $\mathcal{L}$, and z occurs in an argument list, then z can be taken away from the list. Instead its index is introduced.

The functions which are primitive recursive can be obtained by omitting clause XIII and the clauses for the list $\mathcal{L}$.

A <u>convergent computation</u> in $\mathcal{L}$ is a tuple $\langle e,a\rangle$ such that $\{e\}^{\mathcal{L}}(a)\downarrow$ (i.e. $\{e\}^{\mathcal{L}}(a) \simeq r$ for some r). To each convergent computation $\langle e,a\rangle$ we associate an ordinal $|\langle e,a\rangle|^{\mathcal{L}}$ = the least μ such that $\exists r(e,a,r) \in \Gamma^{\mu+1}$. Sometimes this ordinal is denoted by $|\{e\}^{\mathcal{L}}(a)|^{\mathcal{L}}$. Let $\kappa^{\mathcal{L}}$ be the closure ordinal of Γ.

Suppose $\{e\}^{\mathcal{L}}(a)\downarrow$. There is a natural way to define the subcomputations of $\langle e,a\rangle$. The set of subcomputations of $\langle e,a\rangle$ is recursive in $\mathcal{L}$ and $\langle e,a\rangle$, uniformly in $\langle e,a\rangle$ when $\{e\}^{\mathcal{L}}(a)\downarrow$. From the subcomputations of $\langle e,a\rangle$ one can construct the <u>computation tree</u> of $\langle e,a\rangle$.

The definition of a subcomputation of $\langle e,a\rangle$ can be extended to arbitrary $\langle e,a\rangle$. The set of subcomputations of $\langle e,a\rangle$ is recursively enumerable in $\mathcal{L}$, $\langle e,a\rangle$, uniformly in $\langle e,a\rangle$.

<u>Theorem 1</u>: For all $e,a : \{e\}^{\mathcal{L}}(a)\downarrow$ iff the computation tree of $\langle e,a\rangle$ is wellfounded.

Computation trees have been studied in [12] and [13].

§ 3 Connection with Kleene recursion in higher types

In this chapter $S = Tp(0) \cup \ldots \cup Tp(n-1)$, where $n > 0$. Then S_ω can be identified with $Tp(1) \times \ldots \times Tp(n)$. Suppose F is an object of type $n+2$. We prove that there is a list $\mathcal{L}$ of relations, functions and functionals such that recursion in F on $Tp(0), \ldots, Tp(n)$ is essentially the same as recursion in $\mathcal{L}$ on $\mathfrak{A}$. The opposite is also true. If $\mathcal{L}$ is a list which contains relations, functions and functionals expressing the type structure of S, then there are objects of $Tp(n+1)$ and $Tp(n+2)$ such that recursion in $\mathcal{L}$ on $\mathfrak{A}$ is essentially the same as recursion in these objects on the types.

§4 Recursion in normal lists on $\mathfrak{A}$

Let E be the functional defined by:

$$E(f) = \begin{cases} 0 & \text{if } \exists x\, f(x) = 0 \\ 1 & \text{if } \forall x\, f(x) \neq 0 \end{cases}$$

where $f \in {}^{I}S$. A list $\mathcal{L}$ is normal if the equality relation on S is recursive in $\mathcal{L}$, and E is weakly recursive in $\mathcal{L}$. A functional F is recursive in $\mathcal{L}$ if there is an index e such that $F(f) = \{e\}^{\mathcal{L},f}(0)$ for all f. F is weakly recursive in $\mathcal{L}$ if there is a primitive recursive function s(e) such that for all e,a: $\{s(e)\}^{\mathcal{L}}(a) \simeq F(\lambda x\{e\}^{\mathcal{L}}(x,a))$. Moreover if $\{s(e)\}^{\mathcal{L}}(a)\downarrow$ then $|\{s(e)\}^{\mathcal{L}}(a)| > |\{e\}^{\mathcal{L}}(x,a)|$ for all x.

Let $C^{\mathcal{L}}$ be the set of convergent computations.

Theorem 2: Let $\mathcal{L}$ be normal. There is a function p which is partial recursive in $\mathcal{L}$, such that $p(x,y)\downarrow$ if $x \in C^{\mathcal{L}}$ or $y \in C^{\mathcal{L}}$, in which case $p(x,y) \simeq 0$ if $|x|^{\mathcal{L}} \leq |y|^{\mathcal{L}}$, $p(x,y) \simeq 1$ if $|x|^{\mathcal{L}} > |y|^{\mathcal{L}}$.

Theorem 3: Let $\mathcal{L}$ be normal. There is a function φ which is partial recursive in $\mathcal{L}$ such that for all e,a: If $\exists n \in N\{e\}^{\mathcal{L}}(n,a)\downarrow$ then $\varphi(e,a)\downarrow$ and $\{e\}^{\mathcal{L}}(\varphi(e,a),a)\downarrow$. Moreover if $\varphi(e,a) \simeq n$ then $\{e\}^{\mathcal{L}}(n,a)\downarrow$.

Theorem 2 corresponds to theorem 6.1 in [13]. Theorem 3 is a corollary of theorem 2.

Suppose Y is a set of elements in ${}^{S}\omega$, indexed by S, i.e. $Y = \{\alpha_r : r \in S\}$. Then all elements in Y can be coded by one element in ${}^{S}\omega$, namely α defined by: $\alpha(r) = \alpha_{(r)_1}((r)_2)$. This is utilized in theorem 4 (corollary 5.2 in [13]).

Theorem 4: There is a relation R which is recursively enumerable in $\mathcal{L}$ such that for all e,a: $\{e\}^{\mathcal{L}}(a)\uparrow \iff \exists\alpha\, R(\alpha,\langle e,a\rangle)$.

<u>Corollary</u>: The relations which are recursively enumerable in $\mathcal{L}$ are not closed under existential quantifiers over S_ω.

<u>Theorem 7</u> (Grilliot's selection theorem): Let $\mathcal{L}$ be normal. Let $B \subseteq S$ be recursively enumerable in $\mathcal{L}$, a ; $B \neq \emptyset$. Then there is a subset B' of B such that B' is recursive in $\mathcal{L}$, a , and $B' \neq \emptyset$. (Theorem 7.2 in [13].)

<u>Corollary</u>: The relations which are recursively enumerable in $\mathcal{L}$ are closed under existential quantifiers ranging over S .

§ 5 Kleene-recursion in normal objects of type $n+2$, $n > 0$

We use the equivalence result from § 3 to see that results about recursion in higher types can be deduced from corresponding results about recursion on $\mathfrak{A}$.

§ 6 Computation theories on $\mathfrak{A}$

A computation theory on $\mathfrak{A}$ is a pair $(\Theta, ||_\Theta)$. Θ is a set of tuples (e,a,r) where $e \in N$, a is a list of objects from I , $r \in S$. $||_\Theta$ is a function from Θ onto some ordinal κ. φ is Θ-computable if $\varphi(a) \simeq r$ iff $(e,a,r) \in \Theta$ for some e . φ is denoted by $\{e\}_\Theta$. For basic definitions and general results about computation theories we refer to Moschovakis [16] or to the survey papers of Fenstad [2], [3]. We emphazise that our computation theories always are single-valued.

Let X be a subset of I such that X is Θ-computable. X is <u>strongly Θ-finite</u> if the partial functional $\mathcal{F}_X$ defined by

$$\mathcal{F}_X(\varphi) = \begin{cases} 0 & \text{if } \exists x \in X \quad \varphi(x) \simeq 0 \\ 1 & \text{if } \forall x \in X \quad \exists r \neq 0 \quad \varphi(x) \simeq r \end{cases}$$

is weakly Θ-computable. X is <u>weakly Θ-finite</u> if the functional F_X defined by

$$F_X(f) = \begin{cases} 0 & \text{if } \exists x \in X \quad f(x) = 0 \\ 1 & \text{if } \forall x \in X \quad \exists r \neq 0 \quad f(x) = r \end{cases}$$

is weakly Θ-computable. ($\mathcal{F}_X$ is defined on partial functions, F_X on total functions.)

Let $C_\Theta = \{\langle e,a\rangle : \{e\}_\Theta(a)\downarrow\}$. Θ is <u>p-normal</u> if there is a Θ-computable function $p(x,y)$ such that $p(x,y)\downarrow$ if $x \in C_\Theta$ or $y \in C_\Theta$, in which case $p(x,y) = 0$ if $|x|_\Theta \leq |y|_\Theta$, $p(x,y) = 1$ if $|x|_\Theta > |y|_\Theta$.

<u>Lemma 14 and 15</u>: If Θ is p-normal, then Θ admits selection operators for numbers. If E is weakly Θ-computable, then I is weakly finite. If $(\Theta, ||_\Theta)$ is recursion in a normal list, then S is strongly finite.

§7 Abstract Kleene theories

In this chapter we introduce the notion of an abstract recursion theory on $Tp(0),\ldots,Tp(n)$, and prove that this is essentially the same as an abstract recursion theory on $\mathfrak{A}$ when $S = Tp(0) \cup \ldots \cup Tp(n-1)$.

§8 Normal computation theories on $\mathfrak{A}$

Θ is <u>normal</u> if the equality relation on S is Θ-computable, I is weakly finite, S is strongly finite, and Θ is p-normal.

Suppose Θ is normal. There are some interesting ordinals associated to Θ:

$$\kappa = \sup\{|x|_\Theta : x \in C_\Theta\}$$

$$\kappa^0 = \sup\{|x|_\Theta : x \in C_\Theta \cap N\}$$

$$\kappa^a = \sup\{|\langle e,a\rangle|_\Theta : \langle e,a\rangle \in C_\Theta\}$$

$$\varkappa^S = \sup\{|x|_\Theta : x \in C_\Theta \cap S\}$$

$$\varkappa^{S,a} = \sup\{|\langle e,a,b\rangle|_\Theta : \langle e,a,b\rangle \in C_\Theta, e,b \in S\}.$$

The order relation between these ordinals is:

$$\varkappa^o \leq \varkappa^a \leq \varkappa^{S,a} < \varkappa .$$

<u>Definition</u>: Let $\sigma \leq \varkappa_\Theta$. Then σ is <u>a-reflecting</u> if for all $e \in N$: $\exists x|\{e\}_\Theta(x,a)| < \sigma \implies \exists x|\{e\}_\Theta(x,a)| < \varkappa^a$.

<u>Lemma 24</u>: Suppose $x \in B \iff \{e\}_\Theta(x,a)\downarrow$. Then i), ii) and iii) are equivalent. i) There is a subset B' of B which is nonempty and Θ-computable in a, ii) $\exists x|\{e\}_\Theta(x,a)| < \varkappa^a$. iii) $\exists x|\{e\}_\Theta(x,a)| < \varkappa^a_r$.

<u>Lemma 25</u>: Let $B \subseteq I$ be Θ-computable. Then i), ii) and iii) are equivalent. i) B is strongly Θ-finite. ii) For all e,a if $\exists x \in B$ $\{e\}_\Theta(x,a)\downarrow$ then $\exists x \in B$ $|\{e\}_\Theta(x,a)| < \varkappa^a$. iii) If C is a nonempty subset of B which is Θ-semicomputable in a then there is a nonempty subset C' of C which is Θ-computable in a.

The notion of reflection was first introduced by Harrington in [7]. The next theorem and the corollary are proved there.

Let a be fixed. Let $P = \{\langle e,b\rangle : \{e\}_\Theta(a,b)\downarrow$, b is a list of objects from S$\}$. $P \subset S$, hence $P \in {}^S\omega$. P is a complete Θ-semicomputable in a subset of S. $\varkappa^{S,a} < \varkappa^{a,P}$.

<u>Theorem 8</u>: $\varkappa^{S,P,a}$ is a-reflecting.

<u>Corollary</u>: Suppose B is a set of subsets of S, such that B is Θ-semicomputable in a, and B contains an element which is Θ-semicomputable in a. Then B contains an element which is Θ-computable in a.

Definitions: $sc(\Theta) = \{X \subseteq I : X$ is Θ-computable$\}$,

$sc(\Theta,a) = \{X \subseteq I : X$ is Θ-computable in $a\}$,

$en(\Theta) = \{X \subseteq I : X$ is Θ-semicomputable$\}$,

$S\text{-}en(\Theta) = \{X \subseteq S : X$ is Θ-semicomputable$\}$.

Theorem 9: Let Θ be normal. Then there is a normal list $\mathcal{L}$ such that $S\text{-}en(\Theta) = S\text{-}en(\mathcal{L})$, and for all $r \in S : sc(\Theta,r) = sc(\mathcal{L},r)$.

This is an abstract version of the plus 2 theorem in Harrington [7] and also of the plus 1 theorem of Sacks [21]. The original plus 2 theorem of Harrington was a reduction result: Starting out with a normal functional G of type $> n+2$ he constructed a functional F of type $n+2$ such that ${}_n en(G) = {}_n en(F)$. His proof uses the fact that $Tp(n)$ is strongly finite in G. Theorem 9 is an improvement in the sense that we start out with a normal computation theory Θ. Hence in the concrete setting of higher types we only assume that $Tp(n)$ is weakly Θ-finite whereas $Tp(n-1)$ is strongly Θ-finite. Thus theorem 9 gives a kind of characterization result. The proof is quite similar to Harrington's proof in [7]. However, some modifications are necessary, and I am grateful to L. Harrington for helpful suggestions in this connection.

In the last part of chapter 8 there is a characterization of those computation theories which are equivalent to recursion in a normal list. We consider computation theories $(\Theta, ||_\Theta)$ on $\mathfrak{A}$ such that the equality relation on S is Θ-computable, E is weakly recursive in Θ, and Θ is p-normal. This is weaker than normality: We do not suppose that S is strongly finite.

Definition: $En(\Theta) = \{\varphi : \varphi$ is Θ-computable$\}$.

Definition: Θ is Mahlo if for all normal lists $\mathcal{L}$:
$\mathcal{L}$ is Θ-computable $\Longrightarrow \exists x(\kappa^x_{\mathcal{L}} < \kappa^x_\Theta)$.

<u>Theorem 10</u>: Let Θ satisfy the properties mentioned above. Then Θ is not Mahlo iff there is a normal Θ-computable list $\mathcal{L}$ such that $\mathrm{En}(\mathcal{L}) = \mathrm{En}(\Theta)$.

References: Thm. 3.2 in [11]. When Θ is a computation theory on ω this is also proved in [8] and [9]. A different characterization theorem has been developed by D. Normann [19], using his imbedding theory in higher types.

§ 9 More about Mahloness

In ordinal recursion the notion of Mahloness is defined in the following way: An ordinal τ is <u>Mahlo</u> if τ is recursively regular, and all normal functions π which are τ-recursive in constants less than τ have a recursively regular fixed point less than τ. (Definition 4.2 (b) in [1].) The purpose of this chapter is to prove that the definition of Mahloness given in § 8 is a natural generalization of the definition above.

To see this let us regard normal computation theories $(\Theta, ||_\Theta)$ with domain ω, i.e. ω is strongly Θ-finite and $||_\Theta$ is a Θ-norm. Then $\kappa_\Theta^x = \kappa_\Theta$ for all $x \in \omega$. The analogue of the notion of Mahloness as defined in § 8 is the following: Θ is <u>Mahlo_1</u> if $\kappa^{\mathcal{L}} < \kappa_\Theta$ for all normal lists $\mathcal{L}$ which are Θ-computable. The notion of Mahloness can also be defined in a way which is more similar to the definition above. In theorem 11 we prove that the two notions of Mahloness are equivalent.

It is the second notion of Mahloness which has been generalized by Dag Normann in his study [19]. From the characterization theorem in § 8 it follows that this notion is in some sense equivalent to our definition in § 8. A direct equivalence proof similar to the proof of theorem 11 has not yet been provided.

§ 10 Calculation of the lengths of some computations

We develop some machinery needed in § 11. One important tool is the notion of a countable recursion structure.

§ 11 Gaps

Let a be a finite list of individuals, and let $\mathcal{L}$ be a list containing the equality relation on S and the functional E. An ordinal τ is a-constructive if there is an index e and a list a' containing only natural numbers and elements from a such that $\{e\}^{\mathcal{L}}(a')\downarrow$ and $|\{e\}^{\mathcal{L}}(a')| = \tau$. It can easily be proved that there are ordinals $< \kappa^a$ which are not a-constructive. An a-gap is a set $[\nu,\nu')$ $(= \{\mu : \nu \leq \mu < \nu'\})$ of ordinals such that no ordinal in the set is a-constructive, ν' is a-constructive, and the a-constructive ordinals are cofinal in ν. In part A we prove some results about the lengths of a-gaps and the number of a-gaps. Among the results are the following ones: The length of an a-gap is a limit ordinal. If $\tau < \kappa^a$ then there are λ^a a-gaps with length $\geq \tau$. The length of the first a-gap is ω. The length of the first a-gap with length $> \omega$ is $\omega \cdot 2$.

In part B we prove that there are a-reflecting ordinals in many of the a-gaps. (It is the same kind of reflection as we studied in § 8.)

In part C we give a characterization of the first ordinal in an a-gap.

§ 12 On Platek: "Foundations of Recursion Theory"

The main purpose of this chapter is to find an indexfree representation of recursion in a list $\mathcal{L}$ of partial functions and partial monotone functionals on an arbitrary domain $\mathfrak{A}$. To obtain this we introduce Platek's indexfree recursion theory $\mathcal{R}_\omega(\mathcal{B})$ in [26]. The objects in the set $\mathcal{R}_\omega(\mathcal{B})$ are hereditarily consistent. Let HC(τ) denote the set of hereditarily consistent objects of type τ over Ob, where Ob is the universe of the computation domain $\mathfrak{A}$. Then $HC(0) = Ob$, $HC(1)$ is the set of partial functions on Ob, and $HC(2)$ can be identified with the set of partial monotone functionals of type 2. It is natural to introduce the hereditarily consistent objects (= HC), as each partial recursive function with one argument is an element of $HC(1)$, and each partial recursive monotone functional is in $HC(2)$. Also the operator $FP^{(2\to2)\to2}$ which to each $\mathcal{F}$ in $HC(2)$ assigns the least fixed point of $\mathcal{F}$ is hereditarily consistent.

In § 12 we define HC and prove some results which are also proved in [26] (correcting a few mistakes). For $\mathcal{B} \subseteq HC$ we define $\mathcal{R}_\omega(\mathcal{B})$ and $\mathcal{R}_1(\mathcal{B})$ as in [26].

<u>Corollary to theorem 27</u>: Let φ be a partial function. Then φ is partial recursive in $\mathcal{L}$ iff $\varphi \in \mathcal{R}_1(\mathcal{B})$, where $\mathcal{B}$ consists of the objects in $\mathcal{B}_0$ and $\mathcal{L}$, $\mathcal{B}_0$ has the following elements: the characteristic function of N, the functions +1, -1, M, K, L, and the number 0. So $\mathcal{R}_1(\mathcal{B})$ is an indexfree recursion theory yielding the functions which are partial recursive in $\mathcal{L}$.

The following result (the reduction theorem) is proved in [26] by λ-calculus: If $\mathcal{B} \subseteq HC^{1+2}$ then $\mathcal{R}_\omega(\mathcal{B})^{1+3} = \mathcal{R}_{1+1}(\mathcal{B})^{1+3}$. This result is an easy consequence of the results which lead up to the corollary above (theorem 26). Finally we give a direct proof of this result (theorem 29). It is direct in the sense that we do not involve recursion theories with indeces as we do in the proof of theorem 26. It turns out that we can weaken the assumption that $\mathcal{B}_0 \subseteq \mathcal{B}$. It suffices to assume the following weaker condition: "$\uparrow \in \mathcal{R}_{1+1}(\mathcal{B})$" where $\uparrow$ is the object "being undefined".

§ 1 THE COMPUTATION DOMAIN

Let S be an infinite set. A <u>coding scheme</u> for S is a quintuple $\mathcal{S} = (N, +1, M, K, L)$ where $N \subseteq S$ is a copy of the natural numbers with successor function $+1$. M is an injection $S \times S \to S$, i.e. for all $r, s, r', s' \in S$: $M(r,s) = M(r',s') \Rightarrow r = r'$ and $s = s'$. K and L are functions $S \to S$ such that for all $r, s \in S$: $K(M(r,s)) = r$, $L(M(r,s)) = s$. In addition N is closed under K, L and M.

There are some functions and a predicate associated to $\mathcal{S}$.

For each natural number n there is an injection $\langle\ \rangle_n : S^n \to S$ defined by:

$$\langle\ \rangle_0 = 0$$

$$\langle r_1, \ldots, r_{n+1} \rangle_{n+1} = M(\langle r_1, \ldots, r_n \rangle_n, r_{n+1}).$$

$\langle\ \rangle$ is an injection $\bigcup_{n<\omega} S^n \to S$ defined by:

$$\langle r_1, \ldots, r_n \rangle = M(n, \langle r_1, \ldots, r_n \rangle_n).$$

The predicate Seq is the image of $\langle\ \rangle$, i.e.

$$\mathrm{Seq}(r) \iff \exists n\, \exists r_1 \ldots r_n \quad r = \langle r_1,,, r_n \rangle.$$

The elements of Seq are called sequences, and there is a function $\mathrm{lh} : S \to N$ which gives the length of sequences:

$$\mathrm{lh}(r) = \begin{cases} 0 & \text{if } \neg\mathrm{Seq}(r) \\ K(r) & \text{if } \mathrm{Seq}(r) \end{cases}$$

$\lambda r i (r)_i$ is a function $S \times N \to S$ such that $(\langle r_1 \ldots r_i \ldots r_n \rangle)_i = r_i$.

$$(r)_i = \begin{cases} 0 & \text{if } \neg(\mathrm{Seq}(r) \text{ and } i \in N \text{ and } i \le \mathrm{lh}(r)) \\ r_i & \text{if } r = \langle r_1 \ldots r_i \ldots r_n \rangle. \end{cases}$$

Let $I = S \cup {}^S\omega$, where ${}^S\omega$ is the set of functions from S to ω. (ω is the set of natural numbers. ω and N will not be distinguished.)

Notations:

Elements in N, ω : e,f,g,h,i,j,k,l,m,n.

Elements in S (subindividuals): r,s.

Elements in ${}^{S}\omega$: $\alpha,\beta,\gamma,\delta$.

Elements in I (individuals): x,y,z,u,v,w.

Finite lists of elements from I: a,b,c,d.

Total functions $I^n \to S$: f,g,h.

Partial functions $I^n \to S$: φ,ψ.

Relations on I^n : R.

Total functionals : F,G.

Partial functionals : $\mathcal{F}$ (in § 8 also G_τ).

Ordinals : $\epsilon,\eta,\kappa,\lambda,\mu,\nu,\xi,\pi,\rho,\sigma,\tau$.

In § 9 the letters π,ρ,σ are reserved for ordinal functions.

Computation theories : $(\Theta,||_\Theta)$, $(\Psi,||_\Psi)$.

<u>Definition</u>: The triple $(I,S,\mathcal{S})$ is called a <u>computation domain</u>. I is the <u>universe</u> of the computation domain. S is the set of <u>subindividuals</u>.

Let $*$ be the injection $S \to {}^{S}\omega$ defined by:

$$r^*(s) = \begin{cases} 0 & \text{if } s = r \\ 1 & \text{"} \; s \neq r \end{cases}$$

Let $^-$ be a function $I \to S$ defined by:

$$x^- = \begin{cases} r & \text{if } x = r^* \\ 0 & \text{if } x \text{ is not in the image of } * \end{cases}$$

Let $\mathcal{S}$ be a coding scheme for S. It is possible to extend the functions M, K, L to I and hence derive a coding scheme for I, for instance in the following way:

$$M(r,\alpha) = \lambda s\, M(M(r^*(s),\alpha(s)),M(0,1))$$

$$M(\alpha,r) = \lambda s\, M(M(\alpha(s),r^*(s)),M(1,0))$$

$$M(\alpha,\beta) = \lambda s\, M(M(\alpha(s),\beta(s)),M(1,1))$$

$M(r,\alpha)$, $M(\alpha,r)$, $M(\alpha,\beta)$ are elements of ${}^{S}\omega$ because N is closed under M. Let

$$K(\alpha) = (\lambda s\, K\circ K(\alpha(s)))^{-} \quad \text{if} \quad L(\alpha(0)) = M(0,1)$$
$$= \lambda s\, K\circ K(\alpha(s)) \quad \text{otherwise}$$

$$L(\alpha) = (\lambda s\, L\circ K(\alpha(s)))^{-} \quad \text{if} \quad L(\alpha(0)) = M(1,0)$$
$$= \lambda s\, L\circ K(\alpha(s)) \quad \text{otherwise}$$

The extended functions M , K , L have the properties:

$$\forall x y x' y' : \ M(x,y) = M(x',y') \Longrightarrow x = x' \text{ and } y = y' ,$$

$$\forall x\, y : \ M(x,y) \in S \Longleftrightarrow x \in S \text{ and } y \in S ,$$

$$\forall x\, y : \ K(M(x,y)) = x , \ L(M(x,y)) = y .$$

Obviously the functions $\langle\ \rangle_n (n \in \omega)$, $\langle\ \rangle$, lh , $\lambda x\, i(x)_i$ and the predicate Seq can be extended to I since they are defined from M , K , L .

§ 2 RECURSION ON $\mathfrak{A}$

Let the triple $(I,S,\mathcal{S})$ be denoted by $\mathfrak{A}$. There is a natural class of functions associated to $\mathfrak{A}$: the class of primitive recursive functions on $\mathfrak{A}$, denoted by PRF.

It is the smallest class of functions containing:

$$f(x,a) = \begin{cases} 0 & \text{if } x \in N \\ 1 & \text{" } x \notin N \end{cases}$$

$$f(x,a) = \begin{cases} 0 & \text{if } x \in S \\ 1 & \text{" } x \notin S \end{cases}$$

$$f(x,a) = \begin{cases} x & \text{if } x \in S \\ 0 & \text{" } x \notin S \end{cases}$$

$$f(x,a) = \begin{cases} x+1 & \text{if } x \in N \\ 0 & \text{" } x \notin N \end{cases}$$

$$f(a) = m \quad (m \in N)$$

$$f(x,y,a) = \begin{cases} M(x,y) & \text{if } x,y \in S \\ 0 & \text{otherwise} \end{cases}$$

$$f(x,a) = \begin{cases} K(x) & \text{if } x \in S \\ 0 & \text{" } x \notin S \end{cases}$$

$$f(x,a) = \begin{cases} L(x) & \text{if } x \in S \\ 0 & \text{" } x \notin S \end{cases}$$

$$f(x,y,a) = \begin{cases} x(y) & \text{if } x \in S_\omega \text{ and } y \in S \\ 0 & \text{otherwise} \end{cases}$$

and with the following closure properties:

if $f,g \in$ PRF then $h \in$ PRF where h is defined by:

i) $h(a) = f(g(a),a)$,

ii) $h(0,a) = f(a)$

$h(n+1,a) = g(h(n,a),a,n)$

$h(x,a) = 0$ if $x \notin N$

iii) $h(a) = f(a')$ where a' is a permutation of the list a.

Let $R_1 \ldots R_k$ be predicates, $f_1 \ldots f_l$ functions with values in S and $F_1 \ldots F_m$ functionals. (A functional is a total function: $T \to \omega$, where T is the set of total functions from I to S.) The class of functions which are primitive recursive in $R_1 \ldots R_k$, $f_1 \ldots f_l$, $F_1 \ldots F_m$ is obtained by adding the following clauses:

$$f(a,b) = \begin{cases} 0 & \text{if } R_i(a) \\ 1 & \text{"} \ \neg R_i(a) \end{cases} \qquad i = 1 \ldots k,$$

$$f(a,b) = f_i(a), \qquad i = 1 \ldots l,$$

$$f(a) = F_i(\lambda x\, g(x,a)), \qquad i = 1 \ldots m,$$

$$h(a) = f(\lambda r\, g(r,a),a).$$

where g has values in N.

With these clauses one can substitute a function for an element in S_ω, i.e. if f,g are primitive recursive in $R_1 \ldots R_k$, $f_1,,,f_l$, $F_1 \ldots F_m$ then so are h where $h(a) = f(\lambda r\, g(r,a),a)$.

Lemma 1: The graphs of the functions $*$, $^-$, of the extended functions $\langle\ \rangle_n$, lh, $\lambda x i (x)_i$, and the extended predicate Seq are primitive recursive in the equality relation on S and the functional E_S defined below.

$$E_S(f) = \begin{cases} 0 & \text{if } \exists s \in S \quad f(s) = 0 \\ 1 & \text{"} \quad \forall s \in S \quad f(s) \neq 0 \end{cases}$$

where $f : I \to S$ is total.

In [12] Kleene has defined the class of partial recursive functions on the pure types. In [15], [16] Moschovakis defined the class of prime computable functions. Here the class of partial recursive functions on

$\mathfrak{A}$ is defined in an analoguous way. A set of computations is defined inductively by the operator Γ which is given below. In the definition of Γ there is one clause for each of the functions and closure properties which defined the primitive recursive functions. In addition there is one clause for diagonalization (clause XIII).

Let $X \subseteq \bigcup_{n<\omega} I^n$. $\Gamma(X)$ is the subset of $\bigcup_{n<\omega} I^n$ defined by:

For all $n \in N$, all lists a of length n

I $(\langle 1,n+1\rangle,x,a,0) \in \Gamma(X)$ if $x \in N$
$(\langle 1,n+1\rangle,x,a,1) \in \Gamma(X)$ if $x \notin N$

II $(\langle 2,n+1\rangle,x,a,0) \in \Gamma(X)$ if $x \in S$
$(\langle 2,n+1\rangle,x,a,1) \in \Gamma(X)$ if $x \notin S$

III $(\langle 3,n+1\rangle,x,a,x) \in \Gamma(X)$ if $x \in S$
$(\langle 3,n+1\rangle,x,a,0) \in \Gamma(X)$ if $x \notin S$

IV $(\langle 4,n+1\rangle,x,a,x+1) \in \Gamma(X)$ if $x \in N$
$(\langle 4,n+1\rangle,x,a,0) \in \Gamma(X)$ if $x \notin N$

V $(\langle 5,n,m\rangle,a,m) \in \Gamma(X)$

VI $(\langle 6,n+2\rangle,x,y,a,M(x,y)) \in \Gamma(X)$ if $x,y \in S$
$(\langle 6,n+2\rangle,x,y,a,0) \in \Gamma(X)$ otherwise

VII $(\langle 7,n+1\rangle,x,a,K(x)) \in \Gamma(X)$ if $x \in S$
$(\langle 7,n+1\rangle,x,a,0) \in \Gamma(X)$ if $x \notin S$

VIII $(\langle 8,n+1\rangle,x,a,L(x)) \in \Gamma(X)$ if $x \in S$
$(\langle 8,n+1\rangle,x,a,0) \in \Gamma(X)$ if $x \notin S$

IX $(\langle 9,n+2\rangle,x,y,a,x(y)) \in \Gamma(X)$ if $x \in {}^{S}\omega$ and $y \in S$,
$(\langle 9,n+2\rangle,x,y,a,0) \in \Gamma(X)$ if $x \notin {}^{S}\omega$ or $y \notin {}^{S}\omega$

X If $\exists y[(e,a,y) \in X$ and $(e',y,a,x) \in X]$,
then $(\langle 10,n,e,e'\rangle,a,x) \in \Gamma(X)$.

XI If $(e,a,x) \in X$, then $(\langle 11,n+1,e,e'\rangle,0,a,x) \in \Gamma(X)$.
If $\exists y[(\langle 11,n+1,e,e'\rangle,m,a,y) \in X$ and $(e',y,m,a,x) \in X]$,

then $(\langle 11,n+1,e,e'\rangle ,m+1,a,x) \in \Gamma(X)$.

XII If $(e,a',x) \in X$, then $(\langle 12,n,e,i\rangle ,a,x) \in \Gamma(X)$, where a' is obtained from a by moving the $i+1$ - st object in a to the front of the list.

XIII If $(e,a,x) \in X$, then $(\langle 13,n+1\rangle ,e,a,x) \in \Gamma(X)$.

Let $\mathcal{L}$ be the list $R_1,,,R_k$, $\varphi_1 \ldots \varphi_l$, $F_1,,,F_m$ where $R_1 \ldots R_k$ are predicates, $\varphi_1 \ldots \varphi_l$ are partial functions with values in S and $F_1 \ldots F_m$ are functionals. The functions which are partial recursive in $\mathcal{L}$ are obtained by adding the following clauses to Γ:

XIV $(\langle 14,j_i+n,i\rangle ,b,a,0) \in \Gamma(X)$ if $R_i(b)$
$(\langle 14,j_i+n,i\rangle ,b,a,1) \in \Gamma(X)$ if $\neg R_i(b)$,
$i = 1 \ldots k$, b has length j_i,

XV $(\langle 15,j_i+n,i\rangle ,b,a,\varphi_i(b)) \in \Gamma(X)$ if $b \in \mathrm{dom}\,\varphi_i$, $i = 1 \ldots l$,

XVI If $\forall x\, \exists y (e,x,a,y) \in X$, then
$(\langle 16,n,e,i\rangle ,a,F_i(f)) \in \Gamma(X)$ where
$f(x) = y \iff (e,x,a,y) \in X$, $i = 1 \ldots m$.

XVII If $\forall x \in S\ \exists y \in N (e,x,a,y) \in X$ and
$(e',z,a,u) \in X$ then
$(\langle 17,n,e,e'\rangle ,a,u) \in \Gamma(X)$, where
$z \in {}^{S}\omega$ is defined by: $z(x) = y$ iff $(e,x,a,y) \in X$.

Let $\Gamma^0 = \emptyset$, $\Gamma^{\nu+1} = \Gamma^{\nu} \cup \Gamma(\Gamma^{\nu})$, $\Gamma^{\lambda} = \bigcup_{\nu<\lambda} \Gamma^{\nu}$ if $\lim \lambda$. $\Gamma^{\infty} = \bigcup_{\nu \in On} \Gamma^{\nu}$ (On = the ordinals). Γ^{∞} is the set inductively defined by Γ.

For $e \in N$ let $\{e\}^{\mathcal{L}}$ be the partial function defined by: $\{e\}^{\mathcal{L}}(a) \simeq x$ iff $(e,a,x) \in \Gamma^{\infty}$. Then $\{e\}^{\mathcal{L}}$ is singlevalued. Let $\mathcal{P}(\mathcal{L}) = \{\{e\}^{\mathcal{L}} : e \in N\}$. $\mathcal{P}(\mathcal{L})$ is the set of functions which are <u>partial recursive in $\mathcal{L}$</u>. If clause XIII is removed from Γ one obtains the class of functions which are primitive recursive in $\mathcal{L}$.

Let $||^{\mathcal{L}} : \Gamma^{\infty} \to On$ be the function defined by

$$|e,a,x|^{\mathcal{L}} = \text{the least } \mu \text{ such that } (e,a,x) \in \Gamma^{\mu+1}$$

If $(e,a,x) \notin \Gamma^{\infty}$ let $|e,a,x| = \varkappa^{\mathcal{L}}$, where $\varkappa^{\mathcal{L}} = \sup\{|e,a,x|^{\mathcal{L}} : (e,a,x) \in \Gamma^{\infty}\}$.

$\varkappa^{\mathcal{L}}$ is a limit ordinal, and $|e,a,x|^{\mathcal{L}} < \varkappa^{\mathcal{L}}$ for all $(e,a,x) \in \Gamma^{\infty}$.

Computations and subcomputations:

A computation is a tuple (e,a,x). It is convergent if $(e,a,x) \in \Gamma^{\infty}$. Otherwise it is divergent. If $(e,a,x) \in \Gamma^{\infty}$ then x is unique, i.e. $(e,a,x) \in \Gamma^{\infty}$ and $(e,a,x') \in \Gamma^{\infty} \Rightarrow x = x'$. Hence there is no ambiguity in denoting the computation by $\langle e,a\rangle$. Sometimes it will be denoted by $\{e\}^{\mathcal{L}}(a)$. (Hence $\{e\}^{\mathcal{L}}(a)$ has a double meaning: it denotes a computation, and also the object x such that $\{e\}^{\mathcal{L}}(a) \simeq x$.) Let $|\{e\}^{\mathcal{L}}(a)|^{\mathcal{L}} = |\langle e,a\rangle|^{\mathcal{L}} = |e,a,x|^{\mathcal{L}}$ where $\{e\}^{\mathcal{L}}(a) \simeq x$. If there is no x such that $\{e\}^{\mathcal{L}}(a) \simeq x$ let $|\{e\}^{\mathcal{L}}(a)|^{\mathcal{L}} = |\langle e,a\rangle|^{\mathcal{L}} = \varkappa^{\mathcal{L}}$.

Let "$\{e\}^{\mathcal{L}}(a)\downarrow$" be an abbreviation for the statement "there is an x such that $\{e\}^{\mathcal{L}}(a) \simeq x$". "$\{e\}^{\mathcal{L}}(a)\uparrow$" is an abbreviation for the negation of this statement.

Suppose $\{e_0\}^{\mathcal{L}}(a)\downarrow$. By looking at the definition of Γ we see that there is an obvious way to define the subcomputations of $\langle e_0,a\rangle$. First we define the immediate subcomputations (i.s.) of $\langle e_0,a\rangle$ by:

i) If $\{e_0\}^{\mathcal{L}}(a)\downarrow$ by one of the clauses I - IX, XIV, XV then there is no i.s. of $\langle e_0,a\rangle$.

ii) If $\{e_0\}^{\mathcal{L}}(a)\downarrow$ by clause X (substitution) then there are two i.s., namely $\langle e,a\rangle$ and $\langle e',\{e\}^{\mathcal{L}}(a),a\rangle$. If $\{e_0\}^{\mathcal{L}}(n,a)\downarrow$ by clause XI (primitive recursion) then $\langle e_0,0,a\rangle$ has one i.s., namely $\langle e,a\rangle$. $\langle e_0,m+1,a\rangle$ has two i.s., namely $\langle e_0,m,a\rangle$ and $\langle e',\{e_0\}^{\mathcal{L}}(m,a),m,a\rangle$.

iii) If $\{e_0\}^{\mathcal{L}}(a)\downarrow$ by the clauses XII or XIII then there is one i.s., namely $\langle e,a'\rangle$, $\langle e,a\rangle$ respectively.

iv) If $\{e_0\}^{\mathcal{L}}(a)\downarrow$ by clause XVI then there is one i.s. for each $x \in I$, namely $\langle e,x,a\rangle$. If $\{e_0\}^{\mathcal{L}}(a)\downarrow$ by clause XVII, then there is one i.s. for each $x \in S$, namely $\langle c,x,a\rangle$.

$\langle e',a'\rangle$ is a <u>subcomputation</u> of $\langle e,a\rangle$ if there is a finite sequence $x_o,x_1 \ldots x_n$ such that $x_o = \langle e,a\rangle$, $x_n = \langle e',a'\rangle$ and for $i = 0,1,\ldots,n-1 : x_{i+1}$ is an i.s. of x_i.

The relation "x is an i.s. of $\langle e_0,a\rangle$" can be defined for arbitrary $\langle e_0,a\rangle$ (the above definition applies only when $\{e_0\}^{\mathcal{L}}(a)\downarrow$). i), iii) and iv) in the definition is changed as follows: "$\{e_0\}^{\mathcal{L}}(a)\downarrow$" is replaced by: "$e_0$ is an index corresponding to one of the clauses ..... and the length of the list a is the same as the number indicated by e_0 (i.e. the length of $a = (e_0)_2$)." ii) is replaced by ii'):

ii') If $e_0 = \langle 10,n,e,e'\rangle$ and the length of a is n then $\langle e,a\rangle$ is an i.s. of $\langle e_0,a\rangle$. If $\{e\}^{\mathcal{L}}(a)\uparrow$ then $\langle e,a\rangle$ is the only i.s. of $\langle e_0,a\rangle$. If $\{e\}^{\mathcal{L}}(a) \simeq y$ then also $\langle e',y,a\rangle$ is an i.s. of $\langle e_0,a\rangle$.

v) If e_0 is not an index, or the length of a is not $(e_0)_2$ then $\langle e_0,a\rangle$ is the only i.s. of $\langle e_0,a\rangle$.

If $\{e_0\}^{\mathcal{L}}(a)\downarrow$ then this definition gives the same i.s. as the previous definition. The notion of a subcomputation can be defined as before. The subcomputations of $\langle e_0,a\rangle$ can be arranged as a tree. At each node in the tree there is a computation. $\langle e_0,a\rangle$ is put at the top node. If $\langle e',a'\rangle$ occurs at a node, then the immediate subcomputations of $\langle e',a'\rangle$ occur at the nodes immediately below. This tree is called the computation tree of $\langle e,a\rangle$. At each node the branching has one of the following forms:

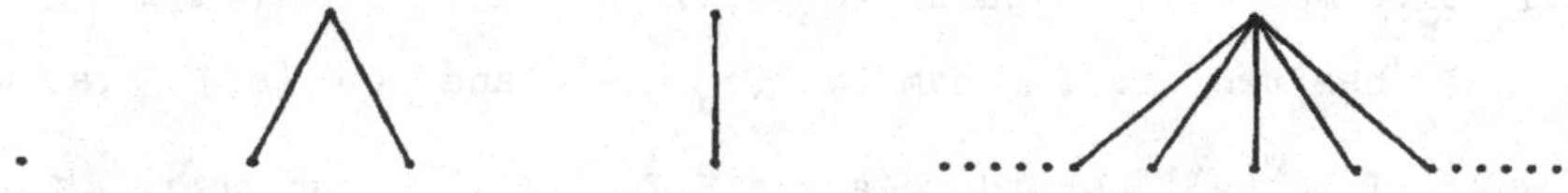

i) ii), ii') ii), ii'), iii), v) iv)

With these conventions the following is true:

Theorem 1: For all e,a : $\{e\}^{\mathcal{L}}(a)\downarrow$ iff the computation tree of $\langle e,a\rangle$ is wellfounded.

Remark 1: It turns out that clause XI (primitive recursion) is superfluous in the presence of clause XIII. It can be replaced by a primitive recursive function. If this function is added and clause XI is omitted the same set of functions will be generated.

Computable functionals and fixpoint theorems:

Let $\mathcal{F}$ be a function : $X \to \omega$ where X is a subset of a cartesian product $P_{m_1} \times P_{m_2} \times \ldots \times P_{m_k} \times I^l$. P_m denotes the set of partial functions : $I^m \to S$. $\mathcal{F}$ is called a partial functional. $\mathcal{F}$ is monotone if $(\varphi_1 \ldots \varphi_k, a) \in \text{dom}\, \mathcal{F}$ and $\varphi_i \subseteq \psi_i$, $i = 1 \ldots k$ $\Rightarrow$ $(\psi_1 \ldots \psi_k, a) \in \text{dom}\, \mathcal{F}$ and $\mathcal{F}(\varphi_1 \ldots \varphi_k, a) \simeq \mathcal{F}(\psi_1 \ldots \psi_k, a)$. $\mathcal{F}$ is partial recursive in $\mathcal{L}$ if there is an index e such that for all $\varphi_1 \ldots \varphi_k, a$:

$$\mathcal{F}(\varphi_1 \ldots \varphi_k, a) \simeq \{e\}^{\mathcal{L}, \varphi_1 \ldots \varphi_k}(a) .$$

$\mathcal{F}$ is weakly partial recursive in $\mathcal{L}$ if there is a primitive recursive function $f(n_1 \ldots n_k)$ such that for all $e_1 \ldots e_k, a_1 \ldots a_k$, where a_i has length n_i, $i = 1 \ldots k$:

$$\mathcal{F}(\varphi_1 \ldots \varphi_k, a) \simeq \{f(n_1 \ldots n_k)\}^{\mathcal{L}}(e_1 \ldots e_k, a_1 \ldots a_k, a)$$

where $\varphi_i = \lambda b\, \{e_i\}^{\mathcal{L}}(b, a_i)$, $i = 1 \ldots k$.

First recursion theorem: Suppose that $\mathcal{F}$ is monotone and weakly partial recursive in $\mathcal{L}$, and that the domain of φ consists of tuples (φ, a) where the length of a is the same as the number of argument places in φ. Then there is a least φ such that for all a : $\mathcal{F}(\varphi, a) \simeq \varphi(a)$, and this φ is partial recursive in $\mathcal{L}$.

<u>Second recursion theorem</u>: $\forall e\, \exists x\, \forall a, \mathcal{L} : \{e\}^{\mathcal{L}}(x,a) \simeq \{x\}^{\mathcal{L}}(a)$.

Let Γ' be the inductive definition which is defined by the clauses I - XII in Γ. Then Γ' generates the class of primitive recursive functions. Let $\{e\}_{PRF}$ denote the primitive recursive function with index e.

<u>Recursion theorem for primitive recursive functions</u>:

Let $f(e,a)$ be primitive recursive. Then there is an e such that for all a : $f(e,a) = \{e\}_{PRF}(a)$.

<u>Lemma 2</u>: Let $\varphi = \lambda b\, \{e\}^{\mathcal{L}}(b,a)$. There is a primitive recursive function f such that for all x, c, y:

$$\{x\}^{\mathcal{L},\varphi}(c) \simeq y \iff \{f(x)\}^{\mathcal{L}}(c,a) \simeq y .$$

An immediate corollary of lemma 2 is

<u>Lemma 3</u>: If $\mathcal{F}$ is partial recursive in $\mathcal{L}$ then $\mathcal{F}$ is weakly partial recursive in $\mathcal{L}$.

Proof of lemma 2: We define a primitive recursive function g by cases. There is one case for each clause in the definition of Γ. $\varphi = \lambda b\, \{e\}^{\mathcal{L}}(b,a)$. Let b have length k and a have length l.

I $x = \langle 1,n+1\rangle$. Let $g(x,t) = \langle 1,n+l+1\rangle$.

Clauses II - IX are treated similarly.

X $x = \langle 10,n,e,e'\rangle$. Let $g(x,t) = \langle 10,n+l,g(e,t),g(e',t)\rangle$.

XIII $x = \langle 13,n+1\rangle$. There is a primitive recursive function h such that for all $t,r,d,\mathcal{L}$: $\{h(t)\}^{\mathcal{L}}(r,d) \simeq \{\{t\}_{PRF}(r)\}^{\mathcal{L}}(d)$. Let $g(x,t) = h(t)$.

XV (the clause for application of φ): $x = \langle 15,k+n,i\rangle$. There is a primitive recursive function s such that for all d of length n : $\{e\}^{\mathcal{L}}(b,a) \simeq \{s(e,n)\}^{\mathcal{L}}(b,\dot{d},a)$ (d is a list of dummy arguments). Let $g(x,t) = s(e,n)$, where n = the length of c minus k.

By the recursion theorem for primitive recursive functions there is a t such that for all x: $g(x,t) = \{t\}_{PRF}(x)$. Let $f(x) = g(x,t)$ for this t. By induction on the length $|\{x\}^{\mathcal{L},\varphi}(c)|^{\mathcal{L},\varphi}$ one can prove: $\{x\}^{\mathcal{L},\varphi}(c) \simeq y \Rightarrow \{f(x)\}^{\mathcal{L}}(c,a) \simeq y$. By induction on $|\{f(x)\}^{\mathcal{L}}(c,a)|$: $\{f(x)\}^{\mathcal{L}}(c,a) \simeq y \Rightarrow \{x\}^{\mathcal{L},\varphi}(c) \simeq y$. □

<u>Remark</u>: The converse of lemma 3 is not true. There are functionals which are weakly partial recursive in $\mathcal{L}$, and which are not partial recursive in $\mathcal{L}$. This can be proved by a cardinality argument as follows: Let

$T_1 = \{\varphi : \exists e,a \quad \varphi = \lambda x\, \{e\}^{\mathcal{L}}(x,a)\}$.

$T_2 = \{\varphi : \varphi$ is a unary partial function $I \to S\}$

$T_3 = T_2 - T_1$. The cardinality of T_1, denoted by $\bar{\bar{T}}_1$, is $\bar{\bar{I}}$.

$\bar{\bar{T}}_2 = 2^{\bar{\bar{I}}}$. $\bar{\bar{T}}_3 = 2^{\bar{\bar{I}}}$.

Let $T_4 = \{\mathcal{F} : \exists e,a \quad \mathcal{F} = \lambda\varphi\, \{e\}_{\mathcal{L},\varphi}(a)\}$, where φ ranges over T_2. $\bar{\bar{T}}_4 = \bar{\bar{I}}$. Let $T_5 = \{\mathcal{F} : \text{dom}\, \mathcal{F} \subseteq T_3\}$. $\bar{\bar{T}}_5 = 2^{(2^{\bar{\bar{I}}})}$. Let $T_6 = \{\mathcal{F} : T_1 \subseteq \text{dom}\, \mathcal{F}$ and $\varphi \in T_1 \Rightarrow \mathcal{F}(\varphi) = 0\}$. $\bar{\bar{T}}_6 = 2^{(2^{\bar{\bar{I}}})}$. Let $\mathcal{F} \in T_6 - T_4$. Then $\mathcal{F}$ is not partial recursive in $\mathcal{L}$, but obviously $\mathcal{F}$ is weakly partial recursive in $\mathcal{L}$.

§ 3 CONNECTION WITH KLEENE RECURSION IN HIGHER TYPES

Recursion in the present setting generalizes recursion in higher types as defined by Kleene. This can be seen as follows.

Let $n > 0$, and let $\epsilon_1 \dots \epsilon_k$ be a list of objects of type $n+1$, $F_1 \dots F_l$ a list of objects of type $n+2$. Let $\mathcal{K}$ denote the set of partial functions φ such that φ is recursive in $\epsilon_1 \dots \epsilon_k, F_1 \dots F_l$ in the sense of Kleene, and the domain of φ is a subset of a cartesion product $U_1 \times \dots \times U_m$, where $U_i = Tp(j)$ for some $j \leq n$ $(i = 1,\dots,m)$.

Let $S = Tp(0) \cup \dots \cup Tp(n-1)$. Let M be a primitive recursive (in the sense of Kleene) pairing function on ω $(= Tp(0))$ such that for all m,n: $M(n,m) > \max(m,n)$. Let K and L be the inverse functions of M. It is possible to extend M, K, L to S in such a way that

i) If $x \in Tp(i)$ $(i < n)$, $y \in Tp(j)$ $(j < n)$ then $M(x,y) \in Tp(k)$, where $k = \max(i,j)$.

ii) For each pair (i,j) such that $i < n$, $j < n$, $\sup(i,j) > 0$, the function f_{ij} is in $\mathcal{K}$, where f_{ij} is defined by:
$f_{ij}(x,y,z) = M(x,y)(z)$, $x \in Tp(i)$, $y \in Tp(j)$, $z \in Tp(k-1)$, $k = \max(i,j)$.

iii) For each pair (i,j) such that $i < n$, $j < n$, $j \leq i$ the function $g_{ij} \in \mathcal{K}$, where $g_{ij}(x,y) = K(x)(y)$ if $K(x) \in Tp(j)$, $= 0$ if $K(x) \notin Tp(j)$, for $x \in Tp(i)$, $y \in Tp(j-1)$, $j > 0$. $g_{io}(x,y) = K(x)$ if $K(x) \in Tp(0)$, $= 0$ if $K(x) \notin Tp(0)$.

iv) Similar conditions for L.

Let $\mathcal{S} = (N, +1, M, K, L)$, $I = S \cup {}^S\omega$. ${}^S\omega$ can be regarded as the product $Tp(1) \times \dots \times Tp(n)$ since $S = Tp(0) \cup \dots \cup Tp(n-1)$. Hence $Tp(n)$ can be regarded as a subset of ${}^S\omega$ via the injection
$Tp(n) \longrightarrow \{0_1\} \times \{0_2\} \times \dots \times \{0_{n-1}\} \times Tp(n)$, where $0_i \in Tp(i)$ is the constant function with value 0.

We want to make a list $\mathcal{L}$ of functions and functionals such that $\mathcal{P}(\mathcal{L})$ is similar to $\mathcal{K}$. For trivial reasons $\mathcal{P}(\mathcal{L})$ is not equal to $\mathcal{K}$. For if $\varphi \in \mathcal{K}$ then

i) the domain of φ is a subset of a fixed cartesian product of types,

ii) the values of φ are in ω.

In $\mathcal{P}(\mathcal{L})$ there are functions which do not satisfy i). If $n > 1$ there are also functions which do not satisfy ii). But this difference between $\mathcal{K}$ and $\mathcal{P}(\mathcal{L})$ is not essential.

Let $\mathcal{L}$ be the list $g_1, g_2, \varepsilon_1' \ldots \varepsilon_k'$, $F_1' \ldots F_l', G, G_1 \ldots G_{n-1}$ where

$$g_1(x) = \begin{cases} Tp(x) & \text{if } x \in S \\ n & \text{" } x \in S_\omega \end{cases}$$

$$g_2(x,y) = \begin{cases} x(y) & \text{if } x \in Tp(1),\ y \in Tp(0) \\ 0 & \text{otherwise} \end{cases}$$

$$\varepsilon_i'(x) = \begin{cases} \varepsilon_i\,(x \mid Tp(n-1)) & \text{if } x \in S_\omega \\ 0 & \text{otherwise} \end{cases}$$

$$i = 1 \ldots k$$

$$G(x,f) = \begin{cases} x(f' \mid Tp(i-2)) & \text{if } x \in Tp(i),\ 2 \leq i < n \\ 0 & \text{otherwise} \end{cases}$$

$$G_i(f) = f' \mid Tp(i-1) \quad \text{for } 1 \leq i < n$$

where $f'(x) = f(x)$ if $f(x) \in \omega$, $= 0$ otherwise,

$$F_i'(f) = F_i\ (f \mid Tp(n)),\quad i = 1, \ldots, l.$$

Lemma 4: There is a primitive recursive function $f(e)$ such that $\{e\}(a) \simeq m$ (in the sense of Kleene) iff $\{f(e)\}^{\mathcal{L}}(a') \simeq m$. (The list a can contain objects of type $\leq n$, and the objects $\varepsilon_1 \ldots \varepsilon_k, F_1 \ldots F_l$. a' is obtained by removing $\varepsilon_1 \ldots \varepsilon_k, F_1 \ldots F_l$.)

Corollary: $\mathcal{K} \subseteq \mathcal{P}(\mathcal{L})$.

Remark: Suppose $\varphi \in \mathcal{K}$. As a member of $\mathcal{K}$ the domain of φ is a subset of a cartesian product $Tp(i_1) \times \ldots \times Tp(i_m)$ $(i_j \leq n)$. As a member of $\mathcal{P}(\mathcal{L})$ the domain of φ is a subset of I^m.

Suppose φ is a partial function such that the domain of φ is a subset of I^m, and the values of φ are in I. φ can be split into components in two ways. First we regard I as $S \cup {}^S\omega$. φ is split into φ' and φ'', where $\varphi'(a) \simeq \varphi(a)$ if $\varphi(a) \in S$, $\simeq 0$ if $\varphi(a) \in {}^S\omega$, undefined if $\varphi(a)$ is undefined. $\varphi''(a) \simeq \varphi(a)$ if $\varphi(a) \in {}^S\omega$, $\simeq 0$ if $\varphi(a) \in S$, undefined if $\varphi(a)$ is undefined, where $0 \in {}^S\omega$ is defined by: $0(r) = 0$ for all $r \in S$. φ'' is partial recursive (primitive recursive) in a list $\mathcal{L}$ if φ''' is, where $\varphi'''(a,y) \simeq \varphi''(a)(y)$ for all $y \in S$. <u>φ is partial recursive (primitive recursive) in $\mathcal{L}$</u> if φ' and φ'' are.

The other way of splitting up φ is natural when we regard I as $Tp(0) \cup Tp(1) \cup \ldots \cup Tp(n-1) \cup (Tp(1) \times Tp(2) \times \ldots \times Tp(n))$. Let $U = X_1 \times \ldots \times X_m$ (φ is m-ary), where X_i is either $Tp(j)$ for some $j < n$, or X_i is $Tp(1) \times \ldots \times Tp(n)$. Then $U \subseteq I^m$. U can be chosen in $(n+1)^m$ different ways. φ can be split into $(n+1)^m$ components, one for each U. Let φ_U be the restriction of φ to U. Each φ_U can be split into φ_{Ui}, $i \leq n$, where $\varphi_{Ui} : U \to Tp(i)$ if $i < n$, $\varphi_{Un} : U \to Tp(1) \times \ldots \times Tp(n)$. If $i < n$ then φ_{Ui} is defined by:

$$\varphi_{Ui}(a) \simeq \begin{cases} \varphi(a) & \text{if } \varphi(a) \in Tp(i) \\ 0_i & \text{if } \varphi(a)\downarrow \text{ and } \varphi(a) \notin Tp(i) \\ \uparrow & \text{if } \varphi(a)\uparrow \end{cases}$$

where $0_i \in Tp(i)$ is defined by: $0_i(x) = 0$ for all $x \in Tp(i-1)$ if $i > 0$. φ_{Un} is defined as φ_{Ui} with $Tp(i)$ replaced by ${}^S\omega$. φ_{Un} can be split into φ_{Unj}, $1 \leq j \leq n$, where $\varphi_{Unj}(a)$ is the j-th component of $\varphi_{Un}(a)$. Hence $\varphi_{Unj} : U \to Tp(j)$. A partial function $\psi : U \to Tp(i)$, $i > 0$ is partial recursive (primitive recursive) in the sense of Kleene if ψ' is, where $\psi' : U \times Tp(i-1) \to Tp(0)$ is defined by: $\psi'(x,y) \simeq \psi(x)(y)$. <u>$\varphi$ is partial recursive (primitive recursive) in the sense of Kleene</u> if all these components are.

<u>Lemma 5</u>: Suppose $\varphi \in \mathcal{P}(\mathcal{L})$. Then φ is partial recursive in $\varepsilon_1 \dots \varepsilon_k, F_1 \dots F_l$ in the sense of Kleene.

<u>Corollary</u>: Let R be a subset of $Tp(n)$. Then R is recursive (recursively enumerable) in $\varepsilon_1 \dots \varepsilon_k, F_1 \dots F_l$ in the sense of Kleene iff R is recursive (recursively enumerable) in $\mathcal{L}$.

<u>Lemma 6</u>: Let $f : I^n \to I$. Then f is primitive recursive in $g_1, g_2, G, G_1 \dots G_{n-1}$ iff f is primitive recursive in the sense of Kleene.

Hence the following definition is meaningful: f is <u>primitive recursive</u> if f is primitive recursive in the sense of Kleene. This definition will be used in the rest of this chapter even if it is not the same as the one given in § 2.

Let $\mathcal{L}$ be the list $g_1, g_2, G, G_1 \dots G_{n-1}, f_1 \dots f_k, F_1 \dots F_l$ where f_i is a total function $I^n \to S$ for $1 \leq i \leq k$, F_i is a total functional with values in S. We want to find a list of objects of type $n+1$ and $n+2$ such that recursion in these objects (in the sense of Kleene) is essentially the same as recursion in $\mathcal{L}$. To construct this list we need two primitive recursive functions p and q between I and $Tp(n)$ such that $p : I \to Tp(n)$, ($Tp(n)$ is regarded as a subset of I), $q : I \to I$, $q(p(x)) = x$ for all x in I. p and q can be constructed from the functions $\langle\langle\ \rangle\rangle$, p_i^j, q_i^j, where $\langle\langle\ \rangle\rangle : \bigcup_{k<\omega} Tp(n)^k \xrightarrow[\text{one-one}]{} Tp(n)$, $p_i^j : Tp(i) \to Tp(j)$, $q_i^j : Tp(j) \to Tp(i)$ for $i < j \leq n$, $q_i^j(p_i^j(\alpha^i)) = \alpha^i$ for all $\alpha \in Tp(i)$. Descriptions of $\langle\langle\ \rangle\rangle$, p_i^j, q_i^j can be found in the works of Kleene [12]. For each k the restriction of $\langle\langle\ \rangle\rangle$ to $Tp(n)^k$ is primitive recursive. So are the functions p_i^j, q_i^j, $\lambda x\, i(x)_i$, lh where $(\langle\langle\alpha_1^n\ldots\alpha_n^n\rangle\rangle)_i = \alpha_i^n$, $lh(\langle\langle\alpha_1^n\ldots\alpha_m^n\rangle\rangle) = m$.

Definition of p: $p(r) = \langle\langle \underline{i}, p_i^n(r)\rangle\rangle$ if $r \in Tp(i)$, $i<n$. $\underline{i} \in Tp(n)$ denotes the constant function with value i. If $x \in {}^{S}\omega = Tp(1)\times\ldots\times Tp(n)$ then $x = (\alpha^1\ldots\alpha^n)$ where $\alpha^i \in Tp(i)$, $i = 1,\ldots n$. Let $p(x) = \langle\langle \underline{n}, p_1^n(\alpha^1), p_2^n(\alpha^2),\ldots, p_{n-1}^n(\alpha^{n-1}), \alpha^n\rangle\rangle$. Let q be defined by: $q(r) = 0$ if $r \in S$. If $x \in {}^{S}\omega$, $x = (\alpha^1,\alpha^2\ldots\alpha^n)$ let

$$q(x) = q_i^n(\alpha^n)_2 \text{ if } lh(\alpha^n) = 2 \text{ and } (\alpha^n)_1(0_{n-1}) = i,$$

$$= (q_1^n((\alpha^n)_2), q_2^n((\alpha^n)_3)\ldots q_{n-1}^n((\alpha^n)_n), (\alpha^n)_{n+1})$$

$$\text{if } lh(\alpha^n) = n+1 \text{ and } (\alpha^n)_1(0_{n-1}) = n$$

$$= 0 \text{ otherwise.}$$

It is routine work to prove that p and q are primitive recursive, and that $q(p(x)) = x$ for all x.

Let $f : I^n \to S$ be one of the functions in the list $\mathcal{L}$. Let $f' \in Tp(n+1)$ be defined by

$$f'(\alpha^n) = [pf(q((\alpha^n)_{1,1}), q((\alpha^n)_{1,2})\ldots q((\alpha^n)_{1,m}))](q_{n-1}^n((\alpha^n)_2)),$$

where $(\alpha^n)_{1,i} = ((\alpha^n)_1)_i$. Then all information about f is contained in f'. This can be seen as follows. Suppose $f(x_1\ldots x_m) = r$. Let $\gamma^n = \langle\langle p(x_1), p(x_2)\ldots p(x_m)\rangle\rangle$, $\alpha^n = \langle\langle\gamma, p_{n-1}^n\beta\rangle\rangle$ for some $\beta \in Tp(n-1)$. Then $f'(\alpha^n) = [pf(x_1\ldots x_m)](\beta) = [p(r)](\beta)$. Hence $p(r) = \lambda\beta^{n-1} f'(\alpha^n)$, and $r = q(\lambda\beta^{n-1} f'(\alpha^n))$.

Let F be one of the functionals in the list $\mathcal{L}$. Let $F' \in Tp(n+2)$ be defined by the following description: Let $\alpha \in Tp(n+1)$. Split α into $(\alpha)_1 = \beta$ and $(\alpha)_2 = \gamma$. Let f be the function from I to S defined by: $f(x) = q\, q_n^{n+1}\beta' p(x)$ if $q\, q_n^{n+1}\beta' p(x) \in S$; $f(x) = 0$ if $q\, q_n^{n+1}\beta' p(x) \in {}^{S}\omega$. β' is the function from $Tp(n)$ to $Tp(n+1)$ defined by: $\beta'(\delta^n) = \lambda\varepsilon^n \beta(\langle\langle\varepsilon,\delta\rangle\rangle)$. As β varies through $Tp(n+1)$, all functions f from I to S will be generated in this way. Suppose $F(f) = y$. Let $F'(\alpha) = p(y)(q_{n-1}^{n+1}(\gamma))$. Then all information about F is contained in F'.

<u>Lemma 7</u>: There is a primitive recursive function $' : \omega \to \omega$ such that for all $e, x_1 \ldots x_m, y$: $\{e\}^{\mathcal{L}}(x_1 \ldots x_m) \simeq y \iff \lambda\beta\{e'\}(p(x_1)\ldots p(x_m),\beta) = p(y)$, where β ranges over $Tp(n-1)$, $\{e'\}$ denotes the e'-th function which is partial recursive in $f_1', \ldots, f_k', F_1', \ldots, F_i'$ in the sense of Kleene.

Proof: We define a primitive recursive function $g : \omega^2 \to \omega$. g is defined by cases. There is one case for each clause in the inductive definition Γ. By the recursion theorem for primitive recursive functions there is a number k such that $g(e,k) = \{k\}_{PRF}(e)$ for all e. Let $e' = g(e,k)$ for this k. It is explained below how to define $g(e,k)$ in the cases diagonalization, application of f_i, and application of F_j.

Diagonalization: $e = \langle 13, h+1 \rangle$. If $\{e\}^{\mathcal{L}}(\hat{e}, a) \simeq x$ then $\{\hat{e}\}^{\mathcal{L}}(a) \simeq x$. There is a primitive recursive function $l(k)$ such that for all $k, \hat{e}, a, \beta$:

$$\{l(k)\}(p(\hat{e}), p(a), \beta) \simeq \{\{k\}_{PRF}(\hat{e})\}(p(a), \beta).$$

Let $g(e,k) = l(k)$.

Application of f_i: $e = \langle 15, m+h, i \rangle$. Let $g(e,k) = e_0$, where e_0 is an index such that $\{e_0\}(p(x_1)\ldots p(x_m), p(a), \beta) \simeq f'(\alpha)$, where $\alpha = \langle\langle \langle\langle p(x_1)\ldots p(x_m)\rangle\rangle, p^n_{n-1}(\beta) \rangle\rangle$.

Application of F_j: $e = \langle 16, h, \hat{e}, j \rangle$. Let $g(e,k)$ be an index such that $\{g(e,k)\}(p(a), \beta) \simeq F'(\alpha)$ where $\alpha \in Tp(n+1)$ is defined such that $(\alpha)_1(\langle\langle \varepsilon, p(x)\rangle\rangle) = (p^{n+1}_n[\lambda\gamma^{n-1}\{g(\hat{e},k)\}(p(x), p(a), \gamma)])(\varepsilon)$ and $(\alpha)_2 = p^{n+1}_{n-1}(\beta^{n-1})$.

By induction on the length of $\{e\}^{\mathcal{L}}(x_1 \ldots x_m)$ it can be proved that $\{e\}^{\mathcal{L}}(x_1 \ldots x_m) \simeq y \implies \lambda\beta\{e'\}(p(x_1)\ldots p(x_m),\beta) = p(y)$. To prove the induction step for the case application of F_j, suppose $\{e\}^{\mathcal{L}}(a) \simeq F_j(\lambda x\{\hat{e}\}^{\mathcal{L}}(x,a)) \simeq y$. Then $|\{\hat{e}\}^{\mathcal{L}}(x,a)| < |\{e\}^{\mathcal{L}}(a)|$ for all x, and

$\{\hat{e}\}^{\mathcal{L}}(x,a) \simeq y' \Rightarrow \lambda\gamma\{\hat{e}'\}(p(x),p(a),\gamma) = p(y')$. $\{e'\}(p(a),\beta) \simeq F'(\alpha)$ by the construction of ', where α is described above. As in the description of F' let f be defined by:

$$\begin{aligned} f(x) &= q\, q_n^{n+1}(\alpha)'_1(p(x)) \\ &= q\, q_n^{n+1}(\lambda\varepsilon(\alpha)_1(\langle\langle\varepsilon,p(x)\rangle\rangle)) \\ &= q\, q_n^{n+1}(\lambda\varepsilon(p_n^{n+1}[\lambda\gamma\{\hat{e}'\}(p(x),p(a),\gamma)])(\varepsilon)) \\ &= q\, q_n^{n+1} p_n^{n+1}[\lambda\gamma\{\hat{e}'\}(p(x),p(a),\gamma)] \\ &= q\,[\lambda\gamma\{\hat{e}'\}(p(x),p(a),\gamma)] \\ &= q\, p(y') \quad \text{by induction hypothesis,} \end{aligned}$$

where $\{\hat{e}\}^{\mathcal{L}}(x,a) \simeq y'$. So $f(x) = y'$. Hence $f = \lambda x\{\hat{e}\}^{\mathcal{L}}(x,a)$.
By the description of F'

$$\begin{aligned} F'(\alpha) &= p(y)(q_{n-1}^{n+1}((\alpha)_2)) \\ &= p(y)(q_{n-1}^{n+1}p_{n-1}^{n+1}(\beta^{n-1})) \\ &= p(y)(\beta^{n-1}). \end{aligned}$$

Hence $\lambda\beta^{n-1}\{e'\}(p(a),\beta) = p(y)$.

By induction on $\min\{|\{e'\}(p(x_1)\ldots p(x_m),\beta)| : \beta \in Tp(n-1)\}$ it can be proved that for all $e',x_1\ldots x_m,y$:

$$\lambda\beta\{e'\}(p(x_1)\ldots p(x_m),\beta) = p(y) \Rightarrow \{e\}^{\mathcal{L}}(x_1\ldots x_m) \simeq y.$$

□

In a similar way one can prove:

Lemma 8: There is a primitive recursive function

$'' : \omega \to \omega$ such that for all $e,\alpha_1\ldots\alpha_m,h$:

$$\{e\}(\alpha_1\ldots\alpha_m) \simeq h \iff \{e''\}^{\mathcal{L}}(\alpha_1\ldots\alpha_m) \simeq h,$$

where α_i ranges over $Tp(j_i)$, $0 \le j_i \le n$, $i = 1\ldots m$.

The purpose of this paper is to reprove some results about recursion in higher types within the framework of chapters 1 and 2. The following should be true: Suppose we have proved a result about recursion on $\mathfrak{A}$. Then there is an easy way to deduce a similar result for recursion in higher types.

When $S = Tp(0) \cup \ldots \cup Tp(n-1)$ we have seen that there is a close correspondence between recursion on $\mathfrak{A}$ and recursion in the sense of Kleene. This correspondence will be utilized in the transition between the two kinds of recursion.

§ 4 RECURSION IN NORMAL LISTS ON $\mathfrak{A}$

In this chapter we will study recursion in lists $\mathcal{L}$ which are normal, i.e. the functional E defined below is weakly recursive in $\mathcal{L}$, the equality relation on S is recursive in $\mathcal{L}$, and $\mathcal{L}$ contains no partial functions.

$$E(f) = \begin{cases} 0 & \text{if } \exists x\, f(x) = 0 \\ 1 & \text{if } \forall x\, \exists y \neq 0 \quad f(x) = y \end{cases}$$

where f is a total function from I to S.

Remark 1: Suppose $S = Tp(0) \cup \ldots \cup Tp(n-1)$. Let $\mathcal{L}$ be a normal list. Construct the list of objects of type $n+1$ and $n+2$ as described in § 3. It can be proved that ${}^{n+2}E$ is weakly recursive in this list, where ${}^{n+2}E \in Tp(n+2)$ is defined by:

$${}^{n+2}E(\alpha^{n+1}) = \begin{cases} 0 & \text{if } \exists \beta \in Tp(n) \quad \alpha(\beta) = 0 \\ 1 & \text{if } \forall \beta \in Tp(n) \quad \alpha(\beta) \neq 0 \end{cases}$$

The opposite is also true. Given a list of objects to type $n+1$ and $n+2$, let $\mathcal{L}$ be the list constructed in § 3. If ${}^{n+2}E$ is weakly recursive in the objects of type $n+1$ and $n+2$, then $\mathcal{L}$ is normal.

In this case the statement "the equality relation on S is recursive in $\mathcal{L}$ " is superfluous in the definition of the notion "normal". The statement can be proved from the fact that E is weakly recursive in $\mathcal{L}$.

Remark 2: In works on higher types the notion "normal" is often defined in a stronger way than here: An object $R \in Tp(n+2)$ is normal if ${}^{n+2}E$ is recursive in F. Here "recursive" is replaced by "weakly recursive". This weaker notion is chosen because it is sufficient in many proofs. (In theorem 5 it is not sufficient.)

Remark 3: If $\mathcal{L}$ is normal then the relations which are recursive in $\mathcal{L}$ are closed under the quantifiers $\forall$ and $\exists$, i.e. if R is recursive in $\mathcal{L}$ then so are $\forall x R$ and $\exists x R$.

Let $\mathcal{L}$ be a list, and let $C^{\mathcal{L}} \subseteq I$ be defined by:

$$C^{\mathcal{L}} = \{\langle e,a\rangle : \{e\}^{\mathcal{L}}(a)\downarrow\}.$$

<u>Theorem 2</u>: Let $\mathcal{L}$ be normal. There is a function p which is partial recursive in $\mathcal{L}$ such that:

$$x \in C^{\mathcal{L}} \text{ or } y \in C^{\mathcal{L}} \iff p(x,y)\downarrow,$$

$$x \in C^{\mathcal{L}} \text{ and } |x|^{\mathcal{L}} \leq |y|^{\mathcal{L}} \Rightarrow p(x,y) = 0,$$

$$|x|^{\mathcal{L}} > |y|^{\mathcal{L}} \Rightarrow p(x,y) = 1.$$

<u>Remark</u>: The index of p can be found in a uniform way. It is a primitive recursive function of t, where t is an index for the primitive recursive function which proves that E is weakly recursive in $\mathcal{L}$.

Proof of theorem 2: Define the partial functional $\mathcal{F}$ be cases. There is one case for each pair of clauses in the inductive definition of Γ. The form of the sequences x and y tells which case we are in. Because there are so many cases (225) only one will be given here: when x corresponds to clause X (substitution) and y to clause XVI (application of F).

Suppose $x = \langle\langle 10,n,e,e'\rangle,a\rangle$ and $y = \langle\langle 16,m,t,1\rangle,b\rangle$. Let p be a partial function, and let

$$\mathcal{F}_1(p,x,y) \simeq E(\lambda z\ p(\langle e,a\rangle,\langle t,z,b\rangle))$$

$$\mathcal{F}_2(p,x,y) \simeq E(\lambda z\ p(\langle e',\{e\}^{\mathcal{L}}(a),a\rangle,\langle t,z,b\rangle))$$

$$\mathcal{F}_3(0,p,x,y) \simeq \mathcal{F}_2(p,x,y), \quad \mathcal{F}_3(1,p,x,y) \simeq 1.$$

Let $\mathcal{F}(p,x,y) \simeq \mathcal{F}_3(\mathcal{F}_1(p,x,y),p,x,y)$. $\mathcal{F}_1$, $\mathcal{F}_2$, $\mathcal{F}_3$ and $\mathcal{F}$ are weakly partial recursive in $\mathcal{L}$ and monotone.

Let p be a solution to the equality $\forall xy(\mathcal{F}(p,x,y) \simeq p(x,y))$.

By induction on $\min\{|x|^{\mathcal{L}}, |y|^{\mathcal{L}}\}$ one can prove:

$$x \in C^{\mathcal{L}} \quad \text{or} \quad y \in C^{\mathcal{L}} \Rightarrow p(x,y)\downarrow ,$$

$$x \in C^{\mathcal{L}} \text{ and } |x|^{\mathcal{L}} \leq |y|^{\mathcal{L}} \Rightarrow p(x,y) = 0 ,$$

$$|x|^{\mathcal{L}} > |y|^{\mathcal{L}} \Rightarrow p(x,y) = 1 .$$

The induction goes as follows: Suppose x and y are as in the case above. If $x \in C^{\mathcal{L}}$ then the immediate subcomputations of x are $\{e\}^{\mathcal{L}}(a)$ and $\{e'\}^{\mathcal{L}}(\{e\}^{\mathcal{L}}(a),a)$, and $|\{e\}^{\mathcal{L}}(a)| < |x|^{\mathcal{L}}$, $|\{e'\}^{\mathcal{L}}(\{e\}^{\mathcal{L}}(a),a)| < |x|^{\mathcal{L}}$. If $y \in C^{\mathcal{L}}$ then $\{t\}^{\mathcal{L}}(z,b)$ is an immediate subcomputation of y for all $z \in I$, and $|\{t\}^{\mathcal{L}}(z,b)| < |y|^{\mathcal{L}}$ for all z. Suppose $x \in C^{\mathcal{L}}$ and $|x|^{\mathcal{L}} \leq |y|^{\mathcal{L}}$. Then by induction hypothesis $p(\langle e,a\rangle,\langle t,z,b\rangle)\downarrow$ and $p(\langle e',\{e\}^{\mathcal{L}}(a),a\rangle,\langle t,z,b\rangle)\downarrow$ for all z. Also

$$\exists z\ |\{e\}^{\mathcal{L}}(a)| \leq |\{t\}^{\mathcal{L}}(z,b)| \quad \text{and}$$

$$\exists z\ |\{e'\}^{\mathcal{L}}(\{e\}^{\mathcal{L}}(a),a)| \leq |\{t\}^{\mathcal{L}}(z,b)| .$$

Hence $\exists z\, p(\langle e,a\rangle,\langle t,z,b\rangle) \simeq 0$ and $\exists z\, p(\langle e',\{e\}^{\mathcal{L}}(a),a\rangle,\langle t,z,b\rangle) \simeq 0$. Hence $\mathcal{F}(p,x,y) \simeq 0$, and $p(x,y) \simeq 0$ since p is a solution to the equality. A similar argument applies when $|x|^{\mathcal{L}} > |y|^{\mathcal{L}}$. This proves the induction.

Let p be the function defined in the theorem. By looking at the definition of $\mathcal{F}$ it can be seen that p is a solution to the equality. By the above induction p is the least solution. By the first recursion theorem p is partial recursive in F.

□

Now we can prove the existence of selection operators for natural numbers.

<u>Theorem 3</u>: Suppose $\mathcal{L}$ is normal. Then there is a function φ which is partial recursive in $\mathcal{L}$ such that for all e,a: If $\exists n \in N$ $\{e\}^{\mathcal{L}}(n,a)\downarrow$ then $\varphi(e,a)\downarrow$, and $\{e\}^{\mathcal{L}}(\varphi(e,a),a)\downarrow$. Moreover if $\varphi(e,a) \simeq n$, then $\{e\}^{\mathcal{L}}(n,a)\downarrow$. The index of φ is a primitive recursive

function of the length of a .

Proof: Let ψ be defined by:

$$\psi(r,n,e,a) \simeq n \quad \text{if} \quad |\{e\}^{\mathcal{L}}(n,a)| \leq |\{r\}^{\mathcal{L}}(n+1,e,a)|$$
$$\simeq \{r\}^{\mathcal{L}}(n+1,e,a) \quad \text{if} \quad |\{e\}^{\mathcal{L}}(n,a)| > |\{r\}^{\mathcal{L}}(n+1,e,a)| .$$

Choose r such that $\psi(r,n,e,a) \simeq \{r\}^{\mathcal{L}}(n,e,a)$ for all n,e,a.
Let $\varphi(e,a) \simeq \{r\}^{\mathcal{L}}(0,e,a)$.

I : $\{e\}^{\mathcal{L}}(n,a)\downarrow \implies \{r\}^{\mathcal{L}}(n,e,a)\downarrow$,

II : $\{r\}^{\mathcal{L}}(n+1,e,a)\downarrow \implies \{r\}^{\mathcal{L}}(n,e,a)\downarrow$.

From I and II it follows:

$$\exists n\{e\}^{\mathcal{L}}(n,a)\downarrow \implies \{r\}^{\mathcal{L}}(0,e,a)\downarrow .$$

Suppose $\{r\}^{\mathcal{L}}(0,e,a) \simeq k$. Want to prove that $\{e\}^{\mathcal{L}}(k,a)\downarrow$. There is an n such that $|\{e\}^{\mathcal{L}}(n,a)| \leq |\{r\}^{\mathcal{L}}(n+1,c,a)|$, for in the opposite case $\{r\}^{\mathcal{L}}(n,e,a) \simeq \{r\}^{\mathcal{L}}(n+1,e,a) \simeq k$ for all n, and $|\{r\}^{\mathcal{L}}(n,e,a)| > |\{r\}^{\mathcal{L}}(n+1,e,a)|$ for all n. Hence we have obtained an infinite descending sequence of ordinals, a contradiction. Let n be the least m such that $|\{e\}^{\mathcal{L}}(m,a)| \leq |\{r\}^{\mathcal{L}}(m+1,e,a)|$. Then $\{e\}^{\mathcal{L}}(n,a)\downarrow$, and $\{r\}^{\mathcal{L}}(n,e,a) \simeq n$. For all $n' < n$ $\{r\}^{\mathcal{L}}(n',e,a) \simeq \{r\}^{\mathcal{L}}(n'+1,e,a) \simeq n$. Hence $\{r\}^{\mathcal{L}}(0,e,a) \simeq n$, i.e. $n = k$, and $\{e\}^{\mathcal{L}}(k,a)\downarrow$.

□

<u>Corollary</u>: If the relations R_1, R_2 are recursively enumerable in $\mathcal{L}$, then so are $\exists n\, R_1$, $R_1 \vee R_2$.

<u>Lemma 9</u>: Suppose that the equality relation on S is recursive in $\mathcal{L}$, and that the functional E_S defined in lemma 1 is weakly recursive in $\mathcal{L}$. Then there is a relation $S(x,y)$ which is recursively enumerable in $\mathcal{L}$ such that if $x \in C^{\mathcal{L}}$ (= the set of convergent computations) then $S(x,y) \iff y$ is an immediate subcomputation of x. The set $\{y:S(x,y)\}$

is recursive in $x, \mathcal{L}$ when $x \in C^{\mathcal{L}}$.

Proof: $S(x,y)$ is defined by cases. There is one case for each clause in Γ. The form of x determines which case is to be applied. We need the functions lh, $\lambda x\, i(x)_i$, and the predicate Seq to decide the form of x; the graphs of these functions and the predicate are primitive recursive in the equality relation on S and the functional E_S (by lemma 1). The only interesting case is when x is a substitution. In the other cases it can be recursively decided whether or not $S(x,y)$ is satisfied. So let $x = \langle\langle 10,n,e,e'\rangle,a\rangle$. Then

$$S(x,y) \iff y = \langle e,a\rangle \text{ or } (\{e\}^{\mathcal{L}}(a)\downarrow \text{ and } y = \langle e',\{e\}^{\mathcal{L}}(a),a\rangle).$$

The relation on the right side is recursively enumerable in $\mathcal{L}$.
If $x \in C^{\mathcal{L}}$ then $\{e\}^{\mathcal{L}}(a)\downarrow$, and the relation is recursive in $x, \mathcal{L}$.

Lemma 10: Suppose $\mathcal{L}$ is normal. If $\{e\}^{\mathcal{L}}(a)\downarrow$ then the computation tree of $\{e\}^{\mathcal{L}}(a)$ is recursive in $\mathcal{L}$, a.

Proof: Let q be the partial function defined by:

$q(x,y)\downarrow$ iff $x \in C^{\mathcal{L}}$,

y is a subcomputation of $x \Rightarrow q(x,y) = 0$,

y is not a subcomputation of $x \Rightarrow q(x,y) = 1$.

Then q is a fixpoint for the monotone weakly $\mathcal{L}$-recursive functional $\mathcal{F}$ defined by:

$$\begin{aligned} \mathcal{F}(q,x,y) \simeq 0 & \text{ if } x \in C^{\mathcal{L}} \text{ and } S(x,y) \\ & \text{ or } \exists z(S(x,z) \text{ and } q(z,y) \simeq 0), \\ \simeq 1 & \text{ if } x \in C^{\mathcal{L}} \text{ and } \neg S(x,y) \\ & \text{ and } \forall z(S(x,z) \Rightarrow q(z,y) \simeq 1). \end{aligned}$$

The quantifiers $\exists$ and $\forall$ can be expressed by E. Hence $\mathcal{F}$ is weakly recursive in $\mathcal{L}$. Suppose q' is a fixpoint for $\mathcal{F}$. By induction on

$|x|^{\mathcal{L}}$ it can be proved that $q(x,y)\downarrow \Rightarrow q(x,y) \simeq q'(x,y)$. Hence q' is an extension of q. Hence q is the least fixpoint of $\mathcal{F}$, and by the first recursion theorem q is partial recursive in $\mathcal{L}$.

If $\{e\}^{\mathcal{L}}(a)\downarrow$, then $\lambda y q(\langle e,a\rangle,y)$ is the characteristic function of the computation tree of $\{e\}^{\mathcal{L}}(a)$. □

Let Y be a set of elements in ${}^{S}\omega$, indexed by S, i.e. $Y = \{\alpha_r : r \in S\}$. Then all elements in Y can be coded by one element in ${}^{S}\omega$, namely α defined by:

$$\alpha(r) = \alpha_{(r)_1}((r)_2).$$

For all $r \in S$ $\lambda s\, \alpha(\langle r,s\rangle) = \alpha_r$. This property will be utilized in the next theorem.

There is a one-one function $**$ from I into ${}^{S}\omega$ and a function $^{--} : I \to I$ such that the graphs of $**$ and $^{--}$ are primitive recursive in the equality relation on S and the functional E_S, and $(x^{**})^{--} = x$ for all x. $**$ and $^{--}$ can for instance be defined by:

$$r^{**} = \langle r^*,\underline{0}\rangle$$

$$\alpha^{**} = \langle \alpha,\underline{1}\rangle$$

where $\underline{0} \in {}^{S}\omega$ and $\underline{1} \in {}^{S}\omega$ denote the constant functions with values 0 and 1 respectively.

$$x^{--} = \begin{cases} (x)_1 & \text{if } (x)_2(0) = 1 \\ (x)_1^- & \text{if } (x)_2(0) = 0\,. \end{cases}$$

<u>Theorem 4</u>: Suppose that the equality relation on S is recursive in $\mathcal{L}$, and that E_S is weakly recursive in $\mathcal{L}$. Then there is a relation R which is recursively enumerable in $\mathcal{L}$ such that for all e,a: $\{e\}^{\mathcal{L}}(a)\uparrow \iff \exists\alpha\, R(\alpha,\langle e,a\rangle)$.

Proof: $\{e\}^{\mathcal{L}}(a)\uparrow \iff$ the computation tree of $\{e\}^{\mathcal{L}}(a)$ is not wellfounded $\iff$

$\exists \alpha_0 \alpha_1 \ldots \alpha_n \ldots [\alpha_0^{--} = \langle e,a \rangle$ and

$\forall i(\alpha_{i+1}^{--}$ is an immediate subcomputation of $\alpha_i^{--})]$

$\iff \exists \alpha_0 \alpha_1 \ldots \alpha_n \ldots (\alpha_0^{--} = \langle e,a \rangle$ and $\forall i\, S(\alpha_i^{--}, \alpha_{i+1}^{--}))$

$\iff \exists \alpha(\alpha[0]^{--} = \langle e,a \rangle$ and $\forall i\, S(\alpha[i]^{--}, \alpha[i+1]^{--}))$

where $\alpha[r] = \lambda s\, \alpha(\langle r,s \rangle)$. Let $R(\alpha\langle e,a \rangle)$

$\iff \alpha[0]^{--} = \langle e,a \rangle$ and $\forall i\, S(\alpha[i]^{--}, \alpha[i+1]^{--})$.

Then R is recursively enumerable in $\mathcal{L}$.

□

Corollary: The relations which are recursively enumerable in $\mathcal{L}$ are not closed under existential quantifiers ranging over S_ω.

Let $P(I)$ denote the set of subsets of I. A relation $R \subseteq P(I) \times I$ is recursive in $\mathcal{L}$ if there is an index e such that $\lambda X x \{e\}^{\mathcal{L},X}(x)$ is the characteristic function of R.

Theorem 5: Suppose that $\mathcal{L}$ is normal and that the functional E is recursive in $\mathcal{L}$. Let $B \subseteq I$. Then B is recursively enumerable in $\mathcal{L}$ iff there is a relation $R \subseteq P(I) \times I$ which is recursive in $\mathcal{L}$ such that for all $x \in I$: $x \in B \iff \exists X$ (X is recursive in $x, \mathcal{L}$ and $(X,x) \in R$).

Proof: Suppose B is recursively enumerable in $\mathcal{L}$. Let e_0 be an index such that for all x: $x \in B \iff \{e_0\}^{\mathcal{L}}(x) \simeq 0$.

We define a relation $S_X(x,y)$ by cases on x. There is one case for each of the clauses in the definition of Γ. Some of the cases are given below: If x is an initial computation then $\neg S_X(x,y)$ for all y. If $x = \langle\langle 10,n,e,e' \rangle,a,z \rangle$ (composition) then $S_X(x,y) \iff x \in X$ and $y \in X$ and $\exists u(y = \langle e,a,u \rangle$ or $y = \langle e',u,a,z \rangle)$. If $x = \langle\langle 16,n,e,i \rangle,a,z \rangle$ (application of the functional F_i) then $S_X(x,y) \iff x \in X$ and $y \in X$ and $\exists x'y'\; y = \langle e,x',a,y' \rangle$.

If x is not of a form which corresponds to a clause in Γ then $S_X(x,y) \iff x \in X$ and $y \in X$ and $x = y$. As a relation of X, x, y S is recursive in $\mathcal{L}$, and $S_X(x,y)$ says that y is an immediate subcomputation of x with respect to X.

Let R be defined by:

$R(X,x) \iff \langle e_0,x,0\rangle \in X$ and

$\forall n\, a\, y\, y'(\langle n,a,y\rangle \in X$ and $\langle n,a,y'\rangle \in X \Rightarrow y = y')$

and $\forall x(x \in X \Rightarrow Seq(x) \wedge lh(x) \geq 2)$

and $\forall \alpha\, \exists i \neg S_X(\alpha[i]^{--}, \alpha[i+1]^{--})$

and $Q(X)$

where Q is the relation which says that if $x \in X$ then x is a convergent computation, and for all y: y is an immediate subcomputation of x iff $S_X(x,y)$.

Obviously all parts of the definition of R except Q are recursive in $\mathcal{L}$. To prove that R is recursive in $\mathcal{L}$ we give instructions how to compute the characteristic function of R. First see if all parts of $R(X,x)$ except $Q(X)$ are satisfied. If not give output 1. If these conditions are satisfied then X can be arranged as a wellfounded tree. Let t_X be the function defined by:

$t_X(x) \simeq 1$ if $x \notin X$

$\simeq 1$ if $x \in X$ and $t_X(y) \simeq 1$ for some y below x in the tree.

In addition there is a case for each clause in the definition of Γ. The cases which correspond to an initial computation, composition and application of F_i are given below:

Initial computation: $t_X(x) \simeq 0$ if $x = \langle n,a,y\rangle$ and $\langle n,a,y\rangle$ is an initial computation and $\{n\}^{\mathcal{L}}(a) \simeq y$. $t_X(x) \simeq 1$ if $\exists y'\{n\}^{\mathcal{L}}(a) \simeq y'$ and $y' \neq y$.

Composition: $x = \langle\langle 10,n,e,e'\rangle,a,z\rangle$.

$$t_X(x) \simeq 0 \text{ if } \exists u[t_X(\langle e,a,u\rangle) \simeq 0 \text{ and } t_X(\langle e',u,a,z\rangle) \simeq 0]$$
$$\simeq 1 \text{ if } \forall u[t_X(\langle e,a,u\rangle) \simeq 1 \text{ or } t_X(\langle e',u,a,z\rangle) \simeq 1]$$

Application of F_i : $x = \langle\langle 16,n,e,i\rangle,a,z\rangle$.

$$t_X(x) \simeq 0 \text{ if } \forall x' \exists y \; t_X(\langle e,x',a,y\rangle) \simeq 0 \text{ and } F_i(\lambda x'\{e\}^{\mathcal{L}}(x',a)) \simeq z$$
$$\simeq 1 \text{ if } \exists x' \forall y \; t_X(\langle e,x',a,y\rangle) \simeq 1 \text{ or } [\forall x' \exists y \; t_X(\langle e,x',a,y\rangle) \simeq 0 \text{ and } \exists z'(F_i(\lambda x'\{e\}^{\mathcal{L}}(x',a)) \simeq z' \text{ and } z' \neq z)]$$

If x does not look like a computation then $t_X(x) \simeq 1$. An index for t_X can be found by the second recursion theorem, hence t_X is partial recursive in $\mathcal{L}$, X, uniformly in X. By induction on the height of x in the tree the following can be proved: $x \in X \Rightarrow t_X(x)$ is defined, and $t_X(x) \simeq 0$ if x is a convergent computation and the part of the tree which lies below x is identical to the computation tree of x, $t_X(x) \simeq 1$ otherwise. Hence $R(X,x)$ iff $t_X(x) \simeq 0$ for all $x \in X$. Hence R is recursive in $\mathcal{L}$.

Suppose $x \in B$. Then $\{e_0\}^{\mathcal{L}}(x) \simeq 0$. Let X be the set of computations in the computation tree of $\{e_0\}^{\mathcal{L}}(x)$ ($\langle e_0,x,0\rangle$ included). Then X is recursive in $x, \mathcal{L}$ by lemma 10, and $R(X,x)$. Hence $\exists X$ (X is recursive in $x, \mathcal{L}$ and $R(X,x)$).
If $\exists X R(X,x)$ then choose X such that $R(X,x)$. X is a set of convergent computations, and $\langle e_0,x,0\rangle \in X$. Hence $\{e_0\}^{\mathcal{L}}(x) \simeq 0$, hence $x \in B$. This proves that

$$x \in B \iff \exists X \text{ (X is recursive in } x, \mathcal{L} \text{ and } R(X,x))$$
$$\iff \exists X \; R(X,x).$$

To prove the other direction of the equivalence in the theorem suppose $x \in B \iff \exists X$ (X is recursive in $x, \mathcal{L}$ and $R(X,x)$) where R is recursive in $\mathcal{L}$. Hence $x \in B \iff \exists m$ ($\lambda y\{m\}^{\mathcal{L}}(x,y)$ is a characteristic function and $R(X,x)$), where X is the set with characteristic function $\lambda y\{m\}^{\mathcal{L}}(x,y)$. Since $\mathcal{L}$ is normal the relations which are recursively enumerable in $\mathcal{L}$ are closed under the quantifier $\exists m$. Hence B is recursively enumerable in $\mathcal{L}$. □

Remark 1: In this proof there are expressions of the form $\forall x(\ldots x,X,\ldots)$ where the expression inside the brackets is recursive in $\mathcal{L}$. The quantifier is expressed by E. This is permitted because E is recursive in $\mathcal{L}$. Weak recursiveness would not suffice.

Theorem 5 can be slightly strengthened. Let $\mathfrak{X}$ be a cartesian product where each factor is one of the following sets: I, the set of functions from I^n into S, the set of functionals. $\mathfrak{A} \subseteq \mathfrak{X}$ is recursively enumerable in $\mathcal{L}$ if there is an index m such that for all $\pi \in \mathfrak{X}$: $\pi \in \mathfrak{A} \iff \{m\}^{\mathcal{L},\pi}(0)\downarrow$.

Theorem 6: Suppose $\mathcal{L}$ is normal and that E is recursive in $\mathcal{L}$. Let $\mathfrak{A} \subseteq \mathfrak{X}$. Then $\mathfrak{A}$ is recursively enumerable in $\mathcal{L}$ iff there is a relation $R \subseteq P(I) \times \mathfrak{X}$ which is recursive in $\mathcal{L}$ such that for all $\pi \in \mathfrak{X}$: $\pi \in \mathfrak{A} \iff \exists X$ (X is recursive in $\mathcal{L}$, π and $(X,\pi) \in R$).
The proof of theorem 6 is a slight modification of the proof of theorem 5.

By theorem 3 there are selection operators for numbers when $\mathcal{L}$ is normal. The next theorem states a similar result for S.

Theorem 7: Suppose $\mathcal{L}$ is normal. There is a function φ which is partial recursive in $\mathcal{L}$ with index $\hat{e}$ such that if $\exists x \in S\,\{e\}^{\mathcal{L}}(x,a)\downarrow$ then $\varphi(\langle e,a\rangle)\downarrow$ and $|\{\hat{e}\}^{\mathcal{L}}(\langle e,a\rangle)| \geq \min\{|\{e\}^{\mathcal{L}}(x,a)| : x \in S\}$. If $\varphi(\langle e,a\rangle)\downarrow$ then $\exists x \in S\,\{e\}^{\mathcal{L}}(x,a)\downarrow$.

Corollary: The relations which are recursively enumerable in $\mathcal{L}$ are closed under existential quantifiers over S.

Proof of theorem 7: The set $\{\langle e,x,a\rangle^{**} : x \in S\}$ is a family of elements in ${}^{S}\omega$ indexed by S. Hence the set can be coded by one element in ${}^{S}\omega$. Call this element α. Then

$$\exists s \in S\ \{e\}^{\mathcal{L}}(s,a)\downarrow \iff \exists s \in S\ \alpha[s]^{--} \in C^{\mathcal{L}},$$

where $C^{\mathcal{L}}$ is the set of convergent computations, and $\alpha[s] = \lambda r\ \alpha(\langle s,r\rangle)$.

Definition: For $\beta \in {}^{S}\omega$ let $\|\beta\| = \min\{|\beta[s]^{--}|^{\mathcal{L}} : s \in S\}$.

Lemma A: There is an index m such that

i) $\|\beta\| < \kappa^{\mathcal{L}} \Rightarrow \{m\}^{\mathcal{L}}(\beta)\downarrow$ and $\|\beta\| \leq |\{m\}^{\mathcal{L}}(\beta)|$,

ii) $\{m\}^{\mathcal{L}}(\beta)\downarrow \Rightarrow \|\beta\| < \kappa^{\mathcal{L}}$.

To prove theorem 7 it is enough to prove lemma A. The index $\hat{e}$ can easily be found from m.

Proof of lemma A: To find m we use the recursion theorem. i) is proved by induction on $\|\beta\|$. Assume as induction hypothesis: $\|\beta\| < \mu \Rightarrow \{m\}^{\mathcal{L}}(\beta)\downarrow$ and $\|\beta\| \leq |\{m\}^{\mathcal{L}}(\beta)|$, for some ordinal μ. Assume $\|\alpha\| = \mu$.

Lemma 9 states that there is a relation $S(x,y)$ which is recursively enumerable in $\mathcal{L}$ such that if $x \in C^{\mathcal{L}}$ then $S(x,y)$ iff y is an immediate subcomputation of x. Let the relation R be defined by:

$R(x,y,w) \iff S(x,y)$ if x is not of the form $\langle\langle 10,n,e,e'\rangle,a\rangle$,

$R(x,y,w) \iff y = \langle e,a\rangle$ or ($w \in C^{\mathcal{L}}$ and $|\{e\}^{\mathcal{L}}(a)| \leq |w|$ and $y = \langle e',\{e\}^{\mathcal{L}}(a),a\rangle$) otherwise.

Let $w \in C^{\mathcal{L}}$. If x is not a substitution then $\{y : R(x,y,w)\}$ is the set of immediate subcomputations of x. If x is a substitution then

$$\{y : R(x,y,w)\} = \begin{cases} \{\langle e,a\rangle\} & \text{if } |w| < |\{e\}^{\mathcal{L}}(a)| \\ \{\langle e,a\rangle, \langle e',\{e\}^{\mathcal{L}}(a),a\rangle\} & \text{otherwise.} \end{cases}$$

For $\sigma < \kappa^{\mathcal{L}}$ let T_σ be the relation defined by:

$$T_\sigma = \{\beta : \forall x \in S \ \ R(\alpha[x]^{--}, \beta[x]^{--}, w)\},$$

where $|w| = \sigma$. Obviously

$$\beta \in T_\sigma \Rightarrow \|\beta\| < \|\alpha\|,$$

$$\sigma < \tau \Rightarrow T_\sigma \subseteq T_\tau .$$

T_σ is recursive in $\mathcal{L}$, α, w where $|w| = \sigma$, since R is recursive in $\mathcal{L}$, w as a relation of x and y when $w \in C^{\mathcal{L}}$.

<u>Lemma B</u>: Let λ be an ordinal such that S is not cofinal in λ (i.e. there is no function $f: S \to \lambda$ such that $\lambda = \sup\{f(x) : x \in S\}$). Let $\{\sigma(\tau) : \tau < \lambda\}$ be an increasing sequence of ordinals bounded above by $\kappa^{\mathcal{L}}$. Then there is an ordinal $\tau' < \lambda$ such that for all τ: $\tau' \leq \tau < \lambda \Rightarrow T_{\sigma(\tau)} = T_{\sigma(\tau')}$.

Proof: To obtain a contradiction suppose
$\forall \tau' < \lambda \ \exists \tau \ (\tau' \leq \tau < \lambda$ and $T_{\sigma(\tau')} \subsetneqq T_{\sigma(\tau)})$.

Take $\tau' < \lambda$. Let τ be minimal such that $\tau' \leq \tau$ and $T_{\sigma(\tau')} \subsetneqq T_{\sigma(\tau)}$. Obviously $\tau' < \tau$. Let $w', w \in C^{\mathcal{L}}$ be chosen such that $|w'| = \sigma(\tau')$. If $\beta \in T_{\sigma(\tau)} - T_{\sigma(\tau')}$ then $\forall x \in S \ R(\alpha[x]^{--}, \beta[x]^{--}, w)$ and $\neg \forall x \in S \ R(\alpha[x]^{--}, \beta[x]^{--}, w')$. Hence $\exists x \in S \ \neg R(\alpha[x]^{--}, \beta[x]^{--}, w')$. If $\neg R(\alpha[x]^{--}, \beta[x]^{--}, w')$ then $\alpha[x]^{--}$ is a substitution $\langle\langle 10, n, e, e'\rangle, a\rangle$, $\beta[x]^{--} = \langle e', \{e\}^{\mathcal{L}}(a), a\rangle$ and $|w'| < |\{e\}^{\mathcal{L}}(a)| \leq |w|$ (because $R(\alpha[x]^{--}, \beta[x]^{--}, w)$). Hence $R(\alpha[x]^{--}, \beta[x]^{--}, w'')$ for all $|w''| \geq |w|$. Let

$$P(\tau') = \{x \in S : \exists \beta \in T_{\sigma(\tau)} - T_{\sigma(\tau')} \ \neg R(\alpha[x]^{--}, \beta[x]^{--}, w')\}.$$

Then $P(\tau')$ is not empty, and $P(\tau') = P(\nu)$ for $\tau' \leq \nu < \tau$ because τ is minimal such that $T_{\sigma(\tau')} \subsetneqq T_{\sigma(\tau)}$. If $\tau \leq \nu$ then $P(\tau')$ and $P(\nu)$ are disjoint, for if $x \in P(\tau')$ then $\alpha[x]^{--} = \langle\langle 10, n, e, e'\rangle, a\rangle$, $\beta[x]^{--} = \langle e', \{e\}^{\mathcal{L}}(a', a)\rangle$ and $R(\alpha[x]^{--}, \beta[x]^{--}, w'')$ for all $w'' \in C^{\mathcal{L}}$

such that $\sigma(\tau) \leq |w''|$. Choose w'' such that $|w''| = \sigma(\nu)$. Then $R(\alpha[x]^{--},\beta[x]^{--},w'')$ since $\sigma(\tau) \leq \sigma(\nu)$. Hence $x \notin P(\nu)$. Let $f : S \to \lambda$ be defined by:

$$f(x) = \begin{cases} \text{the least } \tau' \text{ such that } x \in P(\tau') \\ \text{if } x \in \bigcup_{\tau<\lambda} P(\tau) \\ 0 \quad \text{otherwise} \end{cases}$$

Then $\sup\{f(x) : x \in S\} = \lambda$, a contradiction. □ (Lemma B)

Let $W \subseteq {}^S 2$ be the set of prewellorderings with domain $\subseteq S$. W is recursive in $\mathcal{L}$. For $\delta \in W$ let $Or(\delta)$ be the length of the prewellordering δ. Let $\lambda = \sup\{Or(\delta) : \delta \in W\}$. Then S is not cofinal in λ.

There is an index m_1 such that $\{m_1\}^{\mathcal{L}}(m,\alpha,w)\downarrow$ if $w \in C^{\mathcal{L}}$ and $\{m\}^{\mathcal{L}}(\beta)\downarrow$ for all $\beta \in T_{|w|}$, and if $w \in C^{\mathcal{L}}$ then $|\{m_1\}^{\mathcal{L}}(m,\alpha,w)| > |\{m\}^{\mathcal{L}}(\beta)|$ for all $\beta \in T_{|w|}$. By induction hypothesis $|\{m_1\}^{\mathcal{L}}(m,\alpha,w)| > \|\beta\|$ for all $\beta \in T_{|w|}$.

By the recursion theorem one can find an index m_2 such that $\{m_2\}^{\mathcal{L}}(m,\alpha,\gamma)\downarrow$ if $\gamma \in W$ and for all $\gamma' \in W$: $Or(\gamma') < Or(\gamma) \Rightarrow \{m_2\}^{\mathcal{L}}(m,\alpha,\gamma')\downarrow$ and $\{m_1\}^{\mathcal{L}}(m,\alpha,\langle m_2,m,\alpha,\gamma'\rangle)\downarrow$. When m_2 is chosen in the natural way the following is true:

$|\{m_2\}^{\mathcal{L}}(m,\alpha,\gamma)| > |\{m_2\}^{\mathcal{L}}(m,\alpha,\gamma')|, |\{m_1\}^{\mathcal{L}}(m,\alpha,\langle m_2,m,\alpha,\gamma'\rangle)|$ for all γ' such that $Or(\gamma') < Or(\gamma)$. Hence by induction hypothesis $|\{m_2\}^{\mathcal{L}}(m,\alpha,\gamma)| > \|\beta\|$ for all $\beta \in T_{|\langle m_2,m,\alpha,\gamma'\rangle|}$ when $Or(\gamma') < Or(\gamma)$.

There is an index m_3 such that $\{m_3\}^{\mathcal{L}}(m,\alpha)\downarrow$ if $\forall\gamma \in W\ \{m_2\}^{\mathcal{L}}(m,\alpha,\gamma)\downarrow$, and $|\{m_3\}^{\mathcal{L}}(m,\alpha)| > |\{m_2\}^{\mathcal{L}}(m,\alpha,\gamma)|$ for $\gamma \in W$.

For $\tau < \lambda$ let

$\sigma(\tau) = \inf\{|\{m_2\}^{\mathcal{L}}(m,\alpha,\gamma)| : \gamma \in W \text{ and } Or(\gamma) = \tau\}$. Then $\{\sigma(\tau) : \tau < \lambda\}$ is a strictly increasing sequence of ordinals bounded above by $|\{m_3\}^{\mathcal{L}}(m,\alpha)|$, and hence by $\kappa^{\mathcal{L}}$. By lemma B there is an ordinal $\tau' < \lambda$ such that $T_{\sigma(\tau')} = T_{\sigma(\tau)}$ when $\tau' \leq \tau < \lambda$.

Let $\sigma = \sup\{\sigma(\tau) : \tau < \lambda\}$.

<u>Claim</u>: $\sigma \geq \|\alpha\|$. (Hence $|\{m_3\}^{\mathcal{L}}(m,\alpha)| \geq \|\alpha\|$.)

To obtain a contradiction suppose $\sigma < \|\alpha\|$.

Let $x \in S$. If $\alpha[x]^{--}$ is a substitution $\langle\langle 10,n,e,e'\rangle,a\rangle$ then either $|\{e\}^{\mathcal{L}}(a)| \leq \sigma(\tau')$ or $\sigma \leq |\{e\}^{\mathcal{L}}(a)|$. For if $\sigma(\tau') < |\{e\}^{\mathcal{L}}(a)| < \sigma$ take a $\beta \in T_{\sigma(\tau')}$. Let $\beta'[y] = \beta[y]$ if $y \neq x$, $\beta'[x] = \langle e',\{e\}^{\mathcal{L}}(a),a\rangle$. Then $\beta' \notin T_{\sigma(\tau')}$. But $\beta' \in T_{\sigma(\tau)}$ for $\sigma(\tau) > |\{e\}^{\mathcal{L}}(a)|$. This is contrary to the fact that $T_{\sigma(\tau)} = T_{\sigma(\tau')}$.

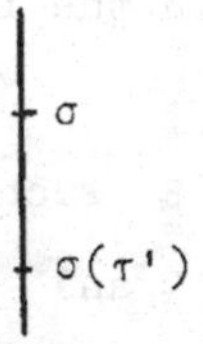

Let β be defined in the following way:
If $\alpha[x]^{--}$ is not a substitution, let $\beta[x]^{--}$ be such that $S(\alpha[x]^{--}, \beta[x]^{--})$ and $|\beta[x]^{--}| \geq \sigma$ (possible by assumption).

If $\alpha[x]^{--}$ is a substitution $\langle\langle 10,n,e,e'\rangle,a\rangle$ let

$$\beta[x]^{--} = \begin{cases} \langle e',\{e\}^{\mathcal{L}}(a),a\rangle & \text{if } |\{e\}^{\mathcal{L}}(a)| \leq \sigma(\tau') \\ \langle e,a\rangle & \text{if } |\{e\}^{\mathcal{L}}(a)| \geq \sigma \end{cases}$$

Then $\|\beta\| \geq \sigma$. $\|\beta\| = \inf\{|\beta[x]^{--}| : x \in S\}$. Obviously $|\beta[x]^{--}| \geq \sigma$ in all cases except for $|\{e\}^{\mathcal{L}}(a)| \leq \sigma(\tau')$. In this case $|\{e'\}^{\mathcal{L}}(\{e\}^{\mathcal{L}}(a),a)| \geq \sigma$ because otherwise $|\alpha[x]^{--}| \leq \sigma$, contrary to the assumption. Hence $|\beta[x]^{--}| \geq \sigma$ also in this case. Hence $\beta \notin T_{\sigma(\tau')}$. For suppose $\beta \in T_{\sigma(\tau')}$. Choose τ such that $\tau' < \tau < \lambda$. Choose $\gamma',\gamma \in W$ such that $Or(\gamma') = \tau'$, $Or(\gamma) = \tau$, $\sigma(\tau') = |\{m_2\}^{\mathcal{L}}(m,\alpha,\gamma')|$, $\sigma(\tau) = |\{m_2\}^{\mathcal{L}}(m,\alpha,\gamma)|$. By the construction of m_2 $|\{m_2\}^{\mathcal{L}}(m,\alpha,\gamma)| > \|\beta'\|$ for all $\beta' \in T_{|\langle m_2,m,\alpha,\gamma'\rangle|}$. Hence $|\{m_2\}^{\mathcal{L}}(m,\alpha,\gamma)| > \|\beta\|$ since $\beta \in T_{\sigma(\tau')} = T_{|\langle m_2,m,\alpha,\gamma'\rangle|}$, contradicting the fact that $|\{m_2\}^{\mathcal{L}}(m,\alpha,\gamma)| = \sigma(\tau) < \sigma \leq \|\beta\|$.

By the construction of β $\beta \in T_{\sigma(\tau')}$, a contradiction. This proves the claim.

By the second recursion theorem there is an m such that

$$\{m_3\}^{\mathcal{L}}(m,\alpha) \simeq \{m\}^{\mathcal{L}}(\alpha) ,$$

$$|\{m\}^{\mathcal{L}}(\alpha)| > |\{m_3\}^{\mathcal{L}}(m,\alpha)|$$

for all α. This m satisfies part i) of lemma A.
By the induction hypothesis mentioned in the beginning of the proof:

$$\|\beta\| < \mu \Rightarrow \{m\}^{\mathcal{L}}(\beta)\downarrow \quad \text{and} \quad \|\beta\| \leq |\{m\}^{\mathcal{L}}(\beta)| .$$

Let $\|\alpha\| = \mu$. By the claim $\|\alpha\| \leq |\{m_3\}^{\mathcal{L}}(m,\alpha)|$. Hence $\|\alpha\| < |\{m\}^{\mathcal{L}}(\alpha)|$.

Part ii) of lemma A can be proved by induction on the length of $|\{m\}^{\mathcal{L}}(\beta)|$. The induction goes as follows.

Suppose $\{m\}^{\mathcal{L}}(\alpha)\downarrow$ and ii) is satisfied for all β such that $|\{m\}^{\mathcal{L}}(\beta)| < |\{m\}^{\mathcal{L}}(\alpha)|$. Since $\{m\}^{\mathcal{L}}(\alpha)\downarrow$, $\{m_3\}^{\mathcal{L}}(m,\alpha)\downarrow$ and $|\{m\}^{\mathcal{L}}(\alpha)| > |\{m_3\}^{\mathcal{L}}(m,\alpha)|$. Also $\{m_2\}^{\mathcal{L}}(m,\alpha,\gamma)\downarrow$ for all $\gamma \in W$, and $|\{m_3\}^{\mathcal{L}}(m,\alpha)| > |\{m_2\}^{\mathcal{L}}(m,\alpha,\gamma)|$ for all $\gamma \in W$.

Let the ordinals $\{\sigma(\tau) : \tau < \lambda\}$ be defined as before. Choose $\tau' < \lambda$ as before. If $\alpha[x]^{--}$ is a substitution, then either $|\{e\}^{\mathcal{L}}(a)| \leq \sigma(\tau')$ or $\sigma \leq |\{e\}^{\mathcal{L}}(a)|$ by the argument in the proof of the claim. To obtain a contradiction suppose $\|\alpha\| = \kappa^{\mathcal{L}}$ (i.e. $\alpha[x]^{--}$ codes a divergent computation for all $x \in S$). Construct β as follows: If $\alpha[x]^{--}$ is not a substitution let $\beta[x]^{--}$ be a divergent subcomputation of $\alpha[x]^{--}$. If $\alpha[x]^{--}$ is a substitution let $\beta[x]^{--}$ be defined as in the proof of the claim. Then by construction $\beta \in T_{\sigma(\tau')}$, and it is easy to check that $|\beta[x]^{--}| \geq \sigma$ for all $x \in S$. Hence $\|\beta\| \geq \sigma$. Choose $\gamma' \in W$ such that $\sigma(\tau') = |\{m_2\}^{\mathcal{L}}(m,\alpha,\gamma')|$. Choose $\gamma \in W$ such that $Or(\gamma') < Or(\gamma)$. Then by the construction of m_2:

$$|\{m_2\}^{\mathcal{L}}(m,\alpha,\gamma)| > |\{m_1\}^{\mathcal{L}}(m,\alpha,\langle m_2,m,\alpha,\gamma'\rangle)| .$$

By the construction of m_1: $|\{m_1\}^{\mathcal{L}}(m,\alpha,\langle m_2,m,\alpha,\gamma'\rangle)| > |\{m\}^{\mathcal{L}}(\beta)|$ since $\beta \in T_{|\langle m_2,m,\alpha,\gamma'\rangle|} = T_{\sigma(\tau')}$. Hence $|\{m\}^{\mathcal{L}}(\alpha)| > |\{m\}^{\mathcal{L}}(\beta)|$. By the induction hypothesis $\|\beta\| < \kappa^{\mathcal{L}}$. By part i) of lemma A: $\|\beta\| \leq |\{m\}^{\mathcal{L}}(\beta)|$. Hence $\|\beta\| < |\{m_2\}^{\mathcal{L}}(m,\alpha,\gamma)|$ for all $\gamma \in W$ such that $Or(\gamma') < Or(\gamma)$. By the definition of $\sigma(\tau)$ $\|\beta\| < \sigma(\tau)$ when $\tau' < \tau < \lambda$. Hence

$\|\beta\| < \sigma$, contradicting the fact that $\|\beta\| \geq \sigma$.

This proves lemma A. □

§ 5 KLEENE RECURSION IN NORMAL OBJECTS OF TYPE $n+2$, $n>0$

Let F be an object to type $n+2$. Let $S = Tp(0) \cup Tp(1) \cup \ldots \cup Tp(n-1)$. Let $\mathcal{S}$ be the coding scheme from § 3. It is proved in § 3 that there is a list $\mathcal{L}$ such that recursion in $\mathcal{L}$ is essentially the same as Kleene recursion in F. There are primitive recursive functions f, g such that

$$\{e\}^F(a) \simeq m \iff \{f(e)\}^{\mathcal{L}}(a) \simeq m,$$

$$|\{e\}^F(a)|^F \leq |\{f(e)\}^{\mathcal{L}}(a)|^{\mathcal{L}},$$

$$\{e\}^{\mathcal{L}}(a) \simeq m \iff \{g(e)\}^F(a) \simeq m,$$

$$|\{e\}^{\mathcal{L}}(a)|^{\mathcal{L}} \leq |\{g(e)\}^F(a)|^F$$

for all e,a,m, where $e,m \in \omega$, a is a list of elements in $\bigcup_{i\leq n} Tp(i)$. (A part of this is stated in lemma 4 and lemma 5.)

As mentioned in § 4 $\mathcal{L}$ is normal iff ${}^{n+2}E$ is weakly recursive in F. In this chapter we will deduce results about Kleene recursion in F from the results in § 4. Let us start with theorem 2.

Let $C^F = \{\langle e,a\rangle : \{e\}^F(a)\downarrow\}$.

Then $\langle e,a\rangle \in C^F \iff \{e\}^F(a)\downarrow \iff \{f(e)\}^{\mathcal{L}}(a)\downarrow$. There is a function $Ord: Tp(n) \to$ Ordinals defined by:

$$Ord(\langle e,a\rangle) = |\{f(e)\}^{\mathcal{L}}(a)|^{\mathcal{L}}$$

Let $x = \langle e,a\rangle$, $y = \langle e',a'\rangle$. Let $p'(x,y) \simeq p(\langle f(e),a\rangle,\langle f(e'),a'\rangle)$ where p is the function from theorem 2. p' is partial recursive in F in the sense of Kleene by lemma 5. This proves

<u>Theorem 2'</u>: Suppose ${}^{n+2}E$ is weakly recursive in F. Then there is a function p' which is partial recursive in F such that

$$x \in C^F \text{ or } y \in C^F \iff p'(x,y)\downarrow,$$

$$x \in C^F \text{ and } Ord(x) \leq Ord(y) \implies p'(x,y) \simeq 0,$$

$$Ord(x) > Ord(y) \implies p'(x,y) \simeq 1.$$

<u>Theorem 3'</u>: Suppose ${}^{n+2}E$ is weakly recursive in F. Then there is a function φ' which is partial recursive in F such that for all e,a: If $\exists n \in N\ \{e\}^F(n,a)\downarrow$ then $\varphi'(e,a)\downarrow$, and $\{e\}^F(\varphi'(e,a),a)\downarrow$. If $\varphi'(e,a) \simeq n$ then $\{e\}^F(n,a)\downarrow$.

Proof: Let $\varphi'(e,a) = \varphi(f(e),a)$, where φ is the selection operator from theorem 3. φ' is partial recursive in F by lemma 5. □

<u>Theorem 4'</u>: Suppose ${}^{n+1}E$ is weakly recursive in F. Then there is a relation R' which is recursively enumerable in F such that for all e,a:

$$\{e\}^F(a)\uparrow \iff \exists\alpha \in Tp(n)\ R'(\alpha,\langle e,a\rangle).$$

Proof: The equality relation on S is recursive in $\mathcal{L}$, and the functional E_S is weakly recursive in $\mathcal{L}$ since ${}^{n+1}E$ is weakly recursive in F. By theorem 4 there is a relation R which is recursively enumerable in $\mathcal{L}$ such that for all n,b:

$$\{n\}^{\mathcal{L}}(b)\uparrow \iff \exists x\ R(x,\langle n,b\rangle). \text{ Hence } \{e\}^F(a)\uparrow \iff \{f(e)\}^{\mathcal{L}}(a)\uparrow$$
$$\iff \exists x\ R(x,\langle f(e),a\rangle).$$

Let p be the function: $I \to Tp(n)$, and $q: I \to I$ the inverse of p, defined in § 3. Then

$$\exists x\ R(x,\langle f(e),a\rangle) \iff \exists\alpha \in Tp(n)\ R(q(\alpha),\langle f(e),a\rangle).$$

Let $R'(\alpha,\langle e,a\rangle) \iff R(q(\alpha),\langle f(e),a\rangle)$. □

<u>Remark</u>: Theorem 4' is a slightly weaker result than corollary 5.2 in [13], where the assumption " ${}^{n+1}E$ is weakly recursive in F " is omitted. Here theorem 4' is a corollary of theorem 4, which is proved for arbitrary sets S. To describe the functions which code elements in I as elements in ${}^S\omega$ we need the equality relation on S and the functional E_S. When working in the type hiearchy the structure is so rich that it

is not necessary to introduce E_S. Hence a better result can be obtained in that case.

Theorem 5': Suppose E is recursive in F. Let $B \subseteq Tp(n)$. B is recursively enumerable in F iff there is a relation $R' \subseteq P(Tp(n)) \times Tp(n)$ which is recursive in F such that for all $x \in Tp(n)$: $x \in B \iff \exists X$ (X is recursive in x, F and $R'(X,x)$).

Proof: When $\mathcal{L}$ is normal $Tp(n)$ is a recursive subset of I. Let $B \subseteq Tp(n)$. B is recursively enumerable in F iff B is recursively enumerable in $\mathcal{L}$. Suppose B is recursively enumerable in F. By theorem 5 there is a relation $R \subseteq P(I) \times I$ which is recursively enumerable in $\mathcal{L}$ such that for all $x \in I$: $x \in B \iff \exists X$ (X is recursive in $x, \mathcal{L}$ and $R(X,x)$). Let R' be defined by:
$R'(Y,y) \iff R(q[Y],y)$, where $Y \subseteq Tp(n), y \in Tp(n)$. Then R' is recursive in $\mathcal{L}$, hence in F. If $q[Y]$ is recursive in $\mathcal{L}, x$ then Y is, and Y is recursive in F, x. Hence $x \in B \iff \exists Y$ (Y is recursive in F, x and $R'(Y,y)$). The other part of theorem 5' can be proved as follows: Suppose $x \in B \iff \exists X$ (X is recursive in x, F and $R'(X,x)$) where R' is recursive in F. Then R' is recursive in $\mathcal{L}$, and X is recursive in x, F iff X is recursive in $\mathcal{L}$. Hence $x \in B \iff \exists X$ (X is recursive in $x, \mathcal{L}$ and $R'(X,x)$). By theorem 5 B is recursively enumerable in $\mathcal{L}$, and hence in F. □

Theorem 6': Suppose ${}^{n+2}E$ is recursive in F. Let $\mathfrak{A} \subseteq Tp(n+2)$. Then $\mathfrak{A}$ is recursively enumerable in F iff there is a relation $R' \subseteq P(Tp(n)) \times Tp(n+2)$ which is recursive in F such that for all $G \in Tp(n+2)$: $G \in \mathfrak{A} \iff \exists X$ (X is recursive in F, G and $R(x,G)$).

Theorem 7': Suppose ${}^{n+2}E$ is weakly recursive in F. Then there is a function φ' which is partial recursive in F with index $\hat{e}'$ such that if $\exists x \in Tp(n-1)\ \{e\}^F(x,a)\downarrow$ then $\varphi'(\langle e,a\rangle)\downarrow$ and $|\{\hat{e}'\}^F(\langle e,a\rangle)| \geq \min\{|\{e\}^F(x,a)| : x \in Tp(n-1)\}$. Moreover if $\varphi'(\langle e,a\rangle)\downarrow$ then $\exists x \in Tp(n-1)\ \{e\}^F(x,a)\downarrow$.

Proof: Let φ be the function from theorem 7 with index $\hat{e}$. If $\exists x \in Tp(n-1)\ \{e\}^F(x,a)\downarrow$ then $\exists x \in Tp(n-1)\ \{f(e)\}^{\mathscr{L}}(x,a)\downarrow$, hence $\exists x \in S\ \{f(e)\}^{\mathscr{L}}(x,a)\downarrow$. Then $\varphi(\langle f(e),a\rangle)\downarrow$, and $|\{\hat{e}\}^{\mathscr{L}}(\langle f(e),a\rangle)| \geq \inf\{|\{f(e)\}^{\mathscr{L}}(x,a)| : x \in S\} \geq \inf\{|\{e\}^F(x,a)| : x \in Tp(n-1)\}$. Let $s = g(\hat{e})$. Then $\{s\}^F(\langle f(e),a\rangle) \simeq \{\hat{e}\}^{\mathscr{L}}(\langle f(e),a\rangle)$, and $|\{s\}^F(\langle f(e),a\rangle)| \geq |\{\hat{e}\}^{\mathscr{L}}(\langle f(e),a\rangle)|$. Choose $\hat{e}'$ such that $\{\hat{e}'\}^F(\langle e,a\rangle) \simeq \{s\}^F(\langle f(e),a\rangle)$ and $|\{\hat{e}'\}^F(\langle e,a\rangle)| \geq |\{s\}^F(\langle f(e),a\rangle)|$. Let $\varphi'(\langle e,a\rangle) \simeq \{\hat{e}'\}^F(\langle e,a\rangle)$. $\square$

§ 6 COMPUTATION THEORIES ON $\mathfrak{A}$

A <u>computation theory</u> on $\mathfrak{A}$ is a pair $(\Theta, ||_\Theta)$ such that the following is true:

Θ is a set of tuples (e,a,r) where $e \in N$, a is a list of individuals, $r \in S$. $||_\Theta$ is a function from Θ onto some ordinal $\varkappa_\Theta$. If $(e,a,r) \in \Theta$ and $(e,a,r') \in \Theta$ then $r = r'$. Let $\{e\}_\Theta$ denote the partial function defined by

$$\{e\}_\Theta(a) \simeq r \iff (e,a,r) \in \Theta$$

φ is <u>Θ-computable</u> if there is an e such that $\varphi = \{e\}_\Theta$, in which case e is said to be an <u>index</u> for φ.

The following functions are Θ-computable: the characteristic functions of N and S, the constant functions $f(a) = n$ $(n \in N)$, M, K, L, i, s, f where

$$i(x,a) = \begin{cases} x & \text{if } x \in S \\ 0 & \text{otherwise} \end{cases}$$

$$s(x) = \begin{cases} x+1 & \text{if } x \in N \\ 0 & \text{otherwise} \end{cases}$$

$$f(x,y) = \begin{cases} x(y) & \text{if } x \in {}^{S}\omega,\ y \in S \\ 0 & \text{otherwise} \end{cases}$$

The following operations are allowed: substitution, primitive recursion, permutations of the list of arguments of a function, adding dummy arguments, substitution of a function for an element in ${}^{S}\omega$, diagonalization, the S^n_m-property is satisfied. To make precise what is meant by "an operation is allowed" let us regard substitution, diagonalization and the S^n_m-property.

Substitution: There is a Θ-computable mapping $g_1(e,f,n)$ such that for all e,f,a,x:

$\{g_1(e,f,n)\}_\Theta(a) \simeq x \iff \exists u[\{e\}_\Theta(a) \simeq u$ and $\{f\}_\Theta(u,a) \simeq x]$

and $|g_1(e,f,n),a,x|_\Theta \geq \sup\{|e,a,u|_\Theta + 1, |f,u,a,x|_\Theta + 1\}$ where $\{e\}_\Theta(a) \simeq u$ and $\{f\}_\Theta(u,a) \simeq x$. The list a has length n.

Diagonalization: There is a Θ-computable mapping $g_2(m,n)$ such that for all e,a,b,x:

$\{g_2(m,n)\}_\Theta(e,a,b) \simeq x \iff \{e\}_\Theta(a) \simeq x$, and

$|g_2(m,n),e,a,b,x|_\Theta \geq |e,a,x|_\Theta + 1$ when $\{e\}_\Theta(a) \simeq x$. The lists a and b have lengths m and n respectively.

S^n_m-property: There is a Θ-computable mapping $g_3(n,m)$ such that $g_3(n,m)$ is an index for a mapping S^n_m with the following property: For all

$e \in N$, $x_1 \ldots x_n \in N$, $y_1 \ldots y_m, z$

$\{S^n_m(e,x_1 \ldots x_n)\}_\Theta(y_1 \ldots y_m) \simeq z \iff \{e\}_\Theta(x_1 \ldots x_n, y_1 \ldots y_m) \simeq z$, and

$|S^n_m(e,x_1 \ldots x_n), y_1 \ldots y_m, z|_\Theta \geq |e,x_1 \ldots x_n, y_1 \ldots y_m, z|_\Theta + 1$, when $\{e\}_\Theta(x_1 \ldots x_n, y_1 \ldots y_m) \simeq z$.

(A mapping is a function which is totally defined.)

This ends the definition of a computation theory on $\mathfrak{A}$.

Let $\mathcal{L}$ be a list of relations, functions and functionals. Let $\Theta = \{(e,a,r) : \{e\}^{\mathcal{L}}(a) \simeq r\}$, and let $|e,a,r|_\Theta = |\{e\}^{\mathcal{L}}(a)|^{\mathcal{L}}$. Then $(\Theta, ||_\Theta)$ is a computation theory on $\mathfrak{A}$.

Some notations and definitions:

Let $(\Theta, ||_\Theta)$ be a computation theory on $\mathfrak{A}$. A <u>computation</u> is a tuple (e,a,r). The computation is <u>convergent</u> if $(e,a,r) \in \Theta$. Otherwise it is <u>divergent</u>. If $(e,a,r) \in \Theta$ then $|e,a,r|_\Theta$ is called the <u>length</u> of the computation (e,a,r). The expression "$\{e\}_\Theta(a)\downarrow$" is an abbreviation for the statement: there is an r such that $\{e\}_\Theta(a) \simeq r$. "$\{e\}_\Theta(a)\uparrow$" is an abbreviation for the negation of this statement. If $\{e\}_\Theta(a)\downarrow$ then there is a unique r such that $\{e\}_\Theta(a) \simeq r$. Hence there is no ambiguity in denoting the computation by $\langle e,a \rangle$. Sometimes it will be denoted by $\{e\}_\Theta(a)$. Hence $\{e\}_\Theta(a)$ has a double meaning. It denotes the object r such that $\{e\}_\Theta(a) \simeq r$, and also the computation (e,a,r). Let $|\{e\}_\Theta(a)|_\Theta = |\langle e,a\rangle|_\Theta = |e,a,r|_\Theta$ where $\{e\}_\Theta(a) \simeq r$. If there is no r such that $\{e\}_\Theta(a) \simeq r$ let $|\{e\}_\Theta(a)|_\Theta = |\langle e,a\rangle|_\Theta = \varkappa_\Theta$, where $\varkappa_\Theta = \sup\{|\{e\}_\Theta(a)|_\Theta : \{e\}_\Theta(a)\downarrow\}$.

The operations (substitution, diagonalization, ...) have the following property: If we start with some computations and perform an operation, then we obtain computations with greater length than the original ones. This corresponds to the intuitive picture of a computation, where the length is a measure of how many operations one must do to obtain a result. The more operations we do, the greater the length will be.

Let $\mathcal{F}$ be a partial monotone functional. $\mathcal{F}$ is <u>weakly Θ-computable</u> if there is a Θ-computable mapping $g(e,n)$ such that for all $e \in N$, $r \in S$, all lists a of length n:
$\{g(e,n)\}_\Theta(a) \simeq r \iff \lambda x\{e\}_\Theta(x,a) \in \mathrm{dom}\,\mathcal{F}$ and $\mathcal{F}(\lambda x\{e\}_\Theta(x,a)) \simeq r$.
Moreover, if $\{g(e,n)\}_\Theta(a) \simeq r$ then there is a subfunction ψ of $\lambda x\{e\}_\Theta(x,a)$ such that $\mathcal{F}(\psi) \simeq r$ and $|\{g(e,n)\}_\Theta(a)|_\Theta > |\{e\}_\Theta(x,a)|_\Theta$ for all $x \in \mathrm{dom}\,\psi$. A <u>Θ-index</u> for $\mathcal{F}$ is an index for the mapping g.

If F is a functional defined on total functions, then by the above definition F is weakly Θ-computable if there is a Θ-computable mapping $g(e,n)$ such that for all $e \in N$, $r \in S$, all lists a of length n: $\{g(e,n)\}_\Theta(a) \simeq r \iff \lambda x\{e\}_\Theta(x,a)$ is total and $F(\lambda x\{e\}_\Theta(x,a)) \simeq r$. If $\{g(e,n)\}_\Theta(a)\downarrow$ then $|\{g(e,n)\}_\Theta(a)|_\Theta > |\{e\}_\Theta(x,a)|_\Theta$ for all $x \in I$.

<u>First recursion theorem</u>: Suppose $\mathcal{F}$ is a partial monotone functional, $\mathcal{F}$ is weakly Θ-computable, and the argument list of $\mathcal{F}$ has the form φ, a, where φ is ranging over k-ary partial functions, and a is ranging over I^k. Then there is a least solution φ to the equality $\forall a[\mathcal{F}(\varphi,a) \simeq \varphi(a)]$, and this least solution is Θ-computable.

<u>Second recursion theorem</u>: $\forall e\, \exists x \forall a\, \{e\}_\Theta(x,a) \simeq \{x\}_\Theta(a)$.

The next two definitions are inspired by Moschovakis: Axioms for Computation Theories - First Draft ([16]). In this paper a subset X of the domain is said to be finite in a computation theory Θ if the relations which are Θ-semicomputable are closed under the quantifiers $\forall x \in X$, $\exists x \in X$ in a uniform way. Below follow two different notions of finitiness. The first one is the same as the one defined in Moschovakis [16].

Let X be a subset of I. X is <u>strongly Θ-finite</u> if the partial functional $\mathcal{F}_X$ defined by

$$\mathcal{F}_X(\varphi) \simeq \begin{cases} 0 & \text{if } \exists x \in X \ \ \varphi(x) \simeq 0 \\ 1 & \text{if } \forall x \in X \ \ \exists r \neq 0 \ \ \varphi(x) \simeq r \end{cases}$$

is weakly Θ-computable. X is <u>weakly Θ-finite</u> if the functional F_X defined by

$$F_X(f) = \begin{cases} 0 & \text{if } \exists x \in X \ \ f(x) = 0 \\ 1 & \text{if } \forall x \in X \ \ \exists r \neq 0 \ \ f(x) = r \end{cases}$$

is weakly Θ-computable. (φ ranges over partial functions, f over functions defined on all of X.) If e is an index for $\mathcal{F}_X$ (F_X) we say that e proves that X is strongly (weakly) finite.

<u>Remark</u>: i) I is weakly Θ-finite $\iff$ E is weakly Θ-computable.

ii) If $X \subseteq I$ is weakly Θ-finite, then the Θ-computable relations on I are closed under the quantifiers $\exists x \in X$, $\forall x \in X$.

iii) If $X \subseteq I$ is strongly Θ-finite, then the Θ-semicomputable relations on I are closed under the quantifiers $\exists x \in X$, $\forall x \in X$.

<u>Lemma 11</u>: If $X \subseteq I$ is strongly Θ-finite, then X is weakly Θ-finite.

<u>Lemma 12</u>: Suppose that the equality relation on I is Θ-computable. Let * denote one of the following two properties: "strongly Θ-finite", "weakly Θ-finite". Suppose $X \subseteq I$ is *. Then i) - iv) are true.

i) X is Θ-computable,

ii) If $Y \subseteq X$ is Θ-computable, then Y is *.

iii) If there is a Θ-computable mapping g such that for all $x \in X$ $g(x)$ is an index which proves that a set Y_x is $*$, then $\bigcup_{x \in X} Y_x$ and $\bigcap_{x \in X} Y_x$ are $*$.

iv) If f is a Θ-computable mapping, then $f[X]$ (the image of X under f) is $*$.

<u>Lemma 13</u>: Let $X = \{x_1 \ldots x_n\}$ where $x_1 \ldots x_n \in I$. Then X is weakly Θ-finite in $x_1 \ldots x_n$. If Θ admits selection operators for natural numbers, then X is also strongly Θ-finite in $x_1 \ldots x_n$.

<u>Definitions</u>: Θ <u>admits selection operators for natural numbers</u> if there is a Θ-computable mapping $g(e,n)$ such that for all $e,n \in N$, all lists a of length n: If $\exists m \in N$ $\{e\}_\Theta(m,a) \simeq 0$, then $\{g(e,n)\}_\Theta(a)\downarrow$ and has a value in N and $\{e\}_\Theta(\{g(e,n)\}_\Theta(a),a) \simeq 0$, and $|\{g(e,n)\}_\Theta(a)| > \inf\{|\{e\}_\Theta(m,a)| : \{e\}_\Theta(m,a) \simeq 0\}$. If $\{g(e,n)\}_\Theta(a) \simeq m$, then $\{e\}_\Theta(m,a) \simeq 0$.

A relation R on I is <u>Θ-computable</u> if the characteristic function of R is Θ-computable. R is <u>Θ-computable in b</u> if there is an index e such that $\lambda a\, \{e\}_\Theta(a,b)$ is the characteristic function of R. R is <u>Θ-semicomputable</u> if there is an index e such that for all a: $R(a) \iff \{e\}_\Theta(a)\downarrow$. R is <u>Θ-semicomputable in b</u> if $R(a) \iff \{e\}_\Theta(a,b)\downarrow$ for some $e \in N$. A partial function φ is <u>Θ-computable in b</u> if $\varphi = \lambda a\, \{e\}_\Theta(a,b)$ for some e. A partial monotone functional $\mathcal{F}$ is weakly <u>Θ-computable in b</u> if there is a Θ-computable mapping $g(e,n)$ such that $\mathcal{F}(\lambda x \{e\}_\Theta(x,a)) \simeq \{g(e,n)\}_\Theta(a,b)$. If $\{g(e,n)\}_\Theta(a,b)\downarrow$ then there is a subfunction ψ of $\lambda x\, \{e\}_\Theta(x,a)$ such that $\mathcal{F}(\psi) \simeq \{g(e,n)\}_\Theta(a,b)$ and $|\{g(e,n)\}_\Theta(a,b)| > |\{e\}_\Theta(x,a)|$ for all $x \in \operatorname{dom}\psi$. A set X is strongly (weakly) Θ-finite in b if $\mathcal{F}_X$ (F_X) is weakly Θ-computable in b.

Let $C_\Theta = \{\langle e,a\rangle : \{e\}_\Theta(a)\downarrow\}$. $||_\Theta$ is a <u>Θ-norm</u> if there is a partial Θ-computable function $p(x,y)$ such that $p(x,y) \simeq 0$ if

$x \in C_\Theta$ and $|x|_\Theta \leq |y|_\Theta$, $p(x,y) \simeq 1$ if $|x|_\Theta > |y|_\Theta$. $(\Theta, ||_\Theta)$ is <u>p-normal</u> if $||_\Theta$ is a Θ-norm.

<u>Lemma 14</u>: If $||_\Theta$ is a Θ-norm, then Θ admits selection operators for natural numbers.

Proof: Same proof as for theorem 3.

<u>Lemma 15</u>: Let $\mathcal{L}$ be a normal list, let Θ denote recursion in $\mathcal{L}$, and let $||_\Theta$ be the natural length function. Then I is weakly Θ-finite, S is strongly Θ-finite and $||_\Theta$ is a Θ-norm.

Proof by theorems 2 and 7.

§ 7 ABSTRACT KLEENE THEORIES

Let F be an object of type $n+2$. Let Θ be the set of tuples (e,a,n) such that $\{e\}^F(a) \simeq n$ in the sense of Kleene. Let $| \, |_\Theta$ denote the natrual length function. Then $\langle \Theta, | \, |_\Theta \rangle$ is not a computation theory in the sense defined in § 6 for some trivial reasons. See the discussion in § 3. The following question arises: Is there a computation theory which is "similar" to $\langle \Theta, | \, |_\Theta \rangle$? This is analoguous to the problems in § 3.

We introduce the notion of an abstract Kleene theory. $\langle \Theta, | \, |_\Theta \rangle$ defined above is an example of such a theory. When $S = Tp(0) \cup \ldots \cup Tp(n-1)$ there is a close connection between Kleene recursion in a list of objects and recursion in a list $\mathcal{L}$ on $\mathfrak{A}$. The motivation for the new notion is to have structures which are related to computation theories on $\mathfrak{A}$ in the same way as Kleene recursion in a list of objects is related to recursion in $\mathcal{L}$.

An abstract Kleene theory on $(Tp(0), \ldots, Tp(n))$ is a pair $\langle \Theta, | \, |_\Theta \rangle$ where Θ is a set of tuples (e,a,n) of length ≥ 2. $e,n \in \omega$, a is a list of objects from $\bigcup_{i \leq n} Tp(i)$. If $(e,a,n) \in \Theta$ and $(e,a,n') \in \Theta$ then $n = n'$. A partial function φ is Θ-computable with index e if for all a,n: $\varphi(a) \simeq n \iff (e,a,n) \in \Theta$. φ is also denoted by $\{e\}_\Theta$. If φ is Θ-computable, then the domain of φ is a subset of a cartesian product $Tp(i_1) \times \ldots \times Tp(i_k)$ where $0 \leq i_j \leq n$ for $1 \leq j \leq k$. The following functions are Θ-computable: $s(n,a) = n+1$, $f(a) = n$, $f(n,a) = n$, $f(n,\alpha,a) = \alpha(n)$ $(\alpha \in Tp(1))$. The following operations are allowed: substitution, primitive recursion, permutations of a list of arguments, diagonalization. The S^n_m-property is satisfied, and the functional F_i is weakly Θ-computable, where $F_i(\alpha,f) = \alpha(\lambda x \in Tp(i) f(x))$, $\alpha \in Tp(i+1)$, $i < n$. The operations satisfy certain ordinal inequalities similar to those given in the definition of a computation theory.

The main difference between a computation theory and an abstract

Kleene theory is that in the latter all computable functions have values in N, and the domain is a subset of a cartesian product of types. In the former theory the computable functions have values in S ($= Tp(0) \cup \ldots \cup Tp(n-1)$), and the domain is a subset of I^k for some k.

<u>Examples</u> of abstract Kleene theories: Let F be an object of type $> n+2$. Let $\Theta = \{(e,a,m) : \{e\}^F(a) \simeq m,\ e,m \in N,\ a$ is a list of objects of type $\leq n\}$. In this case the "natural" Kleene theory is $\{(e,a,m) : \{e\}^F(a) \simeq m,\ a$ is a list of objects of type $\leq k\}$, where the type of F is $k+2 > n+2$. For the latter theory there is a natural length function $||^F$. A length function $||_\Theta$ for Θ can be constructed from $||^F$. If F is a normal object, then $||_\Theta$ is a Θ-norm, and $Tp(i)$ is strongly Θ-finite for $0 \leq i \leq n$.

Let S be the object of type $n+3$ defined by

$$S(e,F) = \begin{cases} 0 & \text{if } \{e\}^F(0)\downarrow \\ 1 & \text{if } \{e\}^F(0)\uparrow \end{cases}$$

where $e \in N$, $F \in Tp(n+2)$. In [7] Harrington constructs a hierarchy for the functions which are recursive in S, F. An abstract Kleene theory can be obtained from this hierarchy. This theory is p-normal, $Tp(n)$ is weakly (not strongly) finite, $Tp(n-1)$ is strongly finite.

Let $S = Tp(0) \cup \ldots \cup Tp(n-1)$. Let $\langle \Theta, ||_\Theta \rangle$ be an abstract Kleene theory on $\mathfrak{A}$. Then there is a computation theory $(\Psi, ||_\Psi)$ which is defined as follows: Let $i = \langle i_1 \ldots i_k \rangle$ where $0 \leq i_j \leq n$ for $j = 1 \ldots k$, and let φ_i be the partial function defined by: $\varphi_i(e,a) \simeq \{e\}_\Theta(a)$, where e ranges over N, a over $Tp(i_1) \times \ldots \times Tp(i_k)$. Then φ_i is Θ-computable. Let $\mathcal{L}$ be the list of the φ_i's and of the functions and functionals which expresses the type structure (i.e. $g_1, g_2, G, G_1, \ldots, G_{n-1}$ in § 3). Let Ψ be the set of convergent computations generated from $\mathcal{L}$. Let $||_\Psi$ be the function defined by: $|e,a,r|_\Psi = 0$ if $(e,a,r) \in \Gamma(\emptyset)$ by some other clause than the clauses for the φ_i's. (Γ is the operator which generates the convergent computations in $\mathcal{L}$).

If $\{e\}_\Theta(a) \simeq n$ then $\varphi_i(e,a) \simeq n$. Let e_i be the index for φ_i. Then $(e_i,e,a,n) \in \Gamma(\emptyset)$. Let $|e_i,e,a,n|_\Psi = |e,a,n|_\Theta$. For other tuples let $|e,a,r|_\Theta$ be the least τ such that $\tau > |e',a',r'|_\Psi$ for all immediate subcomputations (e',a',r') of (e,a,r). Then $(\Psi,||_\Psi)$ is a computation theory.

<u>Lemma 16</u>: Let $\langle\Theta,||_\Theta\rangle$ and $(\Psi,||_\Psi)$ be as above. Then there is a Θ-computable mapping f such that for all e,a,m :

$$\{e\}_\Psi(a) \simeq m \iff \{f(e)\}_\Theta(p(a)) \simeq m \ ;$$

if $\{e\}_\Psi(a) \simeq m$ and (e',a',m') is a subcomputation of (e,a,m) then $|\{f(e')\}_\Theta(p(a'))|_\Theta < |\{f(e)\}_\Theta(p(a))|_\Theta$.

There is a mapping g which is primitive recursive such that for all e,a,m :

$$\{e\}_\Theta(a) \simeq m \iff \{g(e)\}_\Psi(a) \simeq m \ ;$$

if $\{e\}_\Theta(a) \simeq m$ and (e',a',m') is a subcomputation of (e,a,m) then $|\{g(e')\}_\Psi(a')|_\Psi < |\{g(e)\}_\Psi(a)|_\Psi$. ($p$ is the imbedding from I into $Tp(n)$ defined in § 3).

<u>Corollary</u>: Let X be a subset of $Tp(i)$, $i \leq n$.

i) X is Θ-semicomputable iff X is Ψ-semicomputable.

ii) X is Θ-computable iff X is Ψ-computable.

iii) X is weakly Θ-finite iff X is weakly Ψ-finite.

iv) X is strongly Θ-finite iff X is strongly Ψ-finite.

<u>Lemma 17</u>: I is strongly (weakly) Ψ-finite iff $Tp(n)$ is strongly (weakly) Θ-finite. S is strongly (weakly) Ψ-finite iff $Tp(n-1)$ is strongly (weakly) Θ-finite.

<u>Definition</u>: $||_\Psi$ is a <u>Ψ-norm</u> if the function $p(x,y)$ which compares

the lengths of computations is Ψ-computable.

<u>Lemma 18</u>: If $||_\Psi$ is a Ψ-norm, then $||_\Theta$ is a Θ-norm. If Tp(n) is weakly Θ-finite and $||_\Theta$ is a Θ-norm then $||_\Psi$ is a Ψ-norm.

<u>Definition</u>: Let $(\Psi, ||_\Psi)$ be a computation theory on $\mathfrak{A}$, and let $\mathcal{L}$ be a list of objects. $\underline{\Psi \sim \mathcal{P}(\mathcal{L})}$ if for all relations R on A : R is Ψ-semicomputable iff R is recursively enumerable in $\mathcal{L}$.

If $\langle \Theta, ||_\Theta \rangle$ is an abstract Kleene theory and $\underline{F}$ a list of objects of type $\leq n+2$ then the relation $\Theta \sim \underline{F}$ is defined in a similar way.

<u>Lemma 19</u>: Let $\langle \Theta, ||_\Theta \rangle$ be an abstract Kleene theory and let $(\Psi, ||_\Psi)$ be the associated computation theory. Then there is a list $\underline{F}$ such that $\Theta \sim \underline{F}$ iff there is a list $\mathcal{L}$ of total functions and functionals including the functions and functionals which describe the type structure, such that $\Psi \sim \mathcal{P}(\mathcal{L})$.

From the preceding lemmas in this chapter one can conclude: Given an abstract Kleene theory $\langle \Theta, ||_\Theta \rangle$ then there is a computation theory $(\Psi, ||_\Psi)$ with almost the same properties as $\langle \Theta, ||_\Theta \rangle$. The converse of this is also true. If $(\Psi, ||_\Psi)$ is a computation theory then there is an abstract Kleene theory $\langle \Theta, ||_\Theta \rangle$ with almost the same properties as $(\Psi, ||_\Psi)$.

§ 8 NORMAL COMPUTATION THEORIES ON $\mathfrak{A}$

This chapter deals with normal computation theories. A normal computation theory is a generalization of recursion in normal lists. The main results are i) a characterization of which normal computation theories are equivalent to recursion in a normal list, and ii) a result which says that the computable relations on I and the semicomputable relations on S in a normal computation theory can always be obtained from a normal list. The latter result is a generalization of the " + 1 " and " + 2 " theorems.

A computation theory $(\Theta, |\,|_\Theta)$ on $\mathfrak{A}$ is <u>normal</u> if

i) the equality relation on S is Θ-computable,

ii) I is weakly Θ-finite and S is strongly Θ-finite,

iii) $(\Theta, |\,|_\Theta)$ is p-normal.

<u>Remark:</u> If F is a normal object of type $n+2$ then the computation theory obtained from F is normal.

Throughout this chapter $(\Theta, |\,|_\Theta)$ will be a normal computation theory. There are som interesting ordinals associated to Θ. Let X be a subset of I including the natural numbers. Let

$\mathrm{Ord}(X) = \{|\{e\}_\Theta(a)|_\Theta : \{e\}_\Theta(a)\downarrow$ and a is a list of elements from X$\}$.

Let

$$\kappa^X = \sup \mathrm{Ord}(X)$$

$$\lambda^X = \text{the order type of } \mathrm{Ord}(X).$$

Particular cases: $X = N$, $N \cup \{x_1 \ldots x_m\}$, S, $S \cup \{x_1 \ldots x_m\}$, I. For these sets X the ordinal κ^X will be denoted by $\kappa^0, \kappa^a, \kappa^S, \kappa^{S,a}, \kappa_\Theta$ respectively, where $a = x_1 \ldots x_m$. Similarly for λ^X.

Let B be a set and let $P \subseteq B \times B$. P is a <u>prewellordering</u> on B if P is a linear preordering, and there are no infinite descending chains in P. The <u>domain of P</u> (dom P) is the set

$\{x : \exists y((x,y) \in P \text{ or } (y,x) \in P)\}$. Let $|P|$ denote the length of the prewellordering.

Lemma 20: i) Let X be a subset of I including N. Then $\varkappa^X$ is the supremum of the lengths of the prewellorderings with domain $\subseteq I$ which are $\circledast$-computable in elements from X. The sumpremum is not attained.

ii) If there are functions M_X, K_X, L_X such that X is closed under these functions, M_X is a pairing function on X with inverse functions K_X and L_X, and $\circledast$-computable in elements from X, then λ^X is the supremum of the lengths of the prewellorderings with domain $\subseteq X$ which are $\circledast$-computable in elements from X. The supremum is not attained.

Proof: Let P be a prewellordering which is $\circledast$-computable in $a = x_1 \dots x_m$, where $x_1 \dots x_m \in X$, and $\operatorname{dom} P \subseteq I$. There is an index e_1 such that $\{e_1\}_\circledast(x,a) \simeq 0$ if $x \in \operatorname{dom} P$, and if $x \in \operatorname{dom} P$, then $|\{e_1\}_\circledast(y,a)| < |\{e_1\}_\circledast(x,a)|$ for all y below x in P. Hence the function $\rho : \operatorname{dom} P \to$ ordinals defined by $\rho(x) = |\{e_1\}_\circledast(x,a)|$ is order-preserving. There is an index e_2 such that $\{e_2\}_\circledast(a)\downarrow$, and $|\{e_1\}_\circledast(x,a)| < |\{e_2\}_\circledast(a)|$ for all $x \in \operatorname{dom} P$. (Let $\{e_2\}_\circledast(a) \simeq E(f)$ where $f(x) \simeq \{e_1\}_\circledast(x,a)$ if $x \in \operatorname{dom} P$, $f(x) \simeq 1$ otherwise.)

To prove i) let P be as above. Then $|\{e_2\}_\circledast(a)| \geq |P|$. Hence $\varkappa^X > |P|$. If $\nu < \varkappa^X$ choose an index e and a list a of elements from X such that $\{e\}_\circledast(a)\downarrow$ and $\nu < |\{e\}_\circledast(a)|$. Let P be the prewellordering defined by: $(x,y) \in P \Longleftrightarrow |x|_\circledast < |y|_\circledast < |\{e\}_\circledast(a)|$. Then P is $\circledast$-computable in a, and $|P| = |\{e\}_\circledast(a)|$. Hence $\nu < |P|$.

To prove ii) let P be a prewellordering which is $\circledast$-computable in a (a is a list from X), and $\operatorname{dom} P \subseteq X$. Then the set $\{|\{e_1\}_\circledast(x,a)| : x \in \operatorname{dom} P\}$ is a subset of $\operatorname{Ord}(X)$ and has order type $|P|$. $|\{e_2\}_\circledast(a)| \in \operatorname{Ord}(X)$, and $|\{e_1\}_\circledast(x,a)| < |\{e_2\}_\circledast(a)|$ for all $x \in \operatorname{dom} P$. The order type of $\operatorname{Ord}(X)$ is λ^X. Hence $|P| < \lambda^X$. Conversely let $\nu < \lambda^X$. Choose an index e and a list a of elements from X such that $\{e\}_\Theta(a)\downarrow$ and the order type of $\{\mu : \mu \in \mathrm{Ord}(X)$ and $\mu < |\{e\}_\Theta(a)|\} = \nu$. Let P' be the prewellordering defined by: $(x,y) \in P' \iff |x|_\Theta \in \mathrm{Ord}(X)$ and $|y|_\Theta \in \mathrm{Ord}(X)$ and $|x|_\Theta < |y|_\Theta < |\{e\}_\Theta(a)|$. Then P' is Θ-computable in a ; and $|P'| = \nu$. $\mathrm{Dom}(P')$ is not necessarily a subset of X since $e',a' \in X$ does not imply that $\langle e',a'\rangle \in X$. We use the functions M_X, L_X, K_X to construct a prewellordering P such that $|P| = |P'|$, P is Θ-computable in elements from X and $\mathrm{dom}\, P \subseteq X$. This can be done as follows: Via the functions M_X, K_X, L_X one can code finite lists of elements in X as one element in X. Call the coding function $\langle\ \rangle_X$. From $\langle e',a'\rangle_X$ one can regain e',a' via decoding functions which are Θ-computable in elements in X. Let P be defined by: $(x,y) \in P \iff (x',y') \in P'$, where x' is obtained from x in the following way: If $x = \langle e',a'\rangle_X$ then $x' = \langle e',a'\rangle$. □

<u>Lemma 21</u>: If X is one of the following sets: N, $N \cup \{x_1 \dots x_n\}$, S, $S \cup \{x_1 \dots x_n\}$ then $\lambda^X < \kappa^X < \kappa_\Theta$. If $a = (x_1 \dots x_n)$ then $\lambda^a \leq \lambda^{S,a} < \kappa^a$.

Proof: Let X be one of the above sets. Then there are functions M_X, K_X, L_X which satisfies the hypothesis of ii) in lemma 20. If P is a prewellordering with domain $\subseteq X$ then P can be regarded as an element in ${}^S\omega$. There is an index e such that if x is a prewellordering with domain $\subseteq X$ then $\{e\}_\Theta(x)\downarrow$ and $|\{e\}_\Theta(x)| \geq$ the length of the prewellordering x. The relation "x is a prewellordering with domain $\subseteq X$" is Θ-computable in b, where $b = 0$ if $X = N$ or $X = S$, $b = a = (x_1 \dots x_n)$ if $X = N \cup \{x_1 \dots x_n\}$ or $X = S \cup \{x_1 \dots x_n\}$. There is an index e' such that $\{e'\}_\Theta(b)\downarrow$ and $|\{e'\}_\Theta(b)| > |\{e\}_\Theta(x)|$ for all prewellorderings x with domain $\subseteq X$. By ii) of lemma 20 λ^X = supremum of the lengths of the prewellorderings with domain $\subseteq X$ which are Θ-computable in elements from X. Hence $\lambda^X \leq |\{e'\}_\Theta(b)|$, hence $\lambda^X < \varkappa^b \leq \varkappa^a \leq \varkappa^X$. This proves that $\lambda^X < \varkappa^X$, and $\lambda^{S,a} < \varkappa^a$ (let $X = S \cup \{x_1 \ldots x_n\}$). To prove that $\varkappa^X < \varkappa_\Theta$ regard the set $\{\langle m,c\rangle : \{m\}_\Theta(c)\downarrow$, c is a list of elements from $X\}$. This set can be regarded as an element α of ${}^S\omega$. An index e'' can be found such that $\{e''\}_\Theta(\alpha)\downarrow$, and $|\{e''\}_\Theta(\alpha)| > |\{m\}_\Theta(c)|$ for all $\langle m,c\rangle$ in the set. Hence $\varkappa^X \leq |\{e''\}_\Theta(\alpha)|$. Hence $\varkappa^X < \varkappa^\alpha \leq \varkappa_\Theta$. □

<u>Reflection</u>: Harrington introduced the notion of reflection in recursion theory in his thesis [7]. In his exposition the notion is defined in the following way: An ordinal σ is a-reflecting if for each Σ_1-formula $\varphi(x)$ of a language $\mathcal{L}$: If $M_\sigma \models \varphi(a)$ then $M_{\varkappa^a} \models \varphi(a)$. M_σ is a structure constructed from the set of computations of length less than σ. The interesting a-reflecting ordinals are those which are greater than $\varkappa^a$. Suppose σ is such an ordinal. Then the following is true: If a Σ_1-sentence $\varphi(a)$ is satisfied in the large model M_σ then it is satisfied in a smaller model. This is a reflecting property.

The notion of reflection can also be defined within the framework of this paper. An approach to the notion can be done by regarding the following problem: Let B be a Θ-semicomputable in a, nonempty subset of I. Is there a subset B' of B such that B' is nonempty and Θ-computable in a ?

This is not true for all B if the relations which are Θ-semicomputable, are not closed under existential quantifiers over I. This can be seen as follows: Let $P(x,a)$ be a Θ-semicomputable relation such that the relation $\exists x P(x,a)$ is not Θ-semicomputable. Let $B_a = \{x : P(x,a)\}$. Then B_a is Θ-semicomputable in a, and $B_a \neq \emptyset$ iff $\exists x P(x,a)$. To obtain a contradiction suppose that there is a nonempty subset B'_a of B_a which is Θ-computable in a for all a such that $B_a \neq \emptyset$. Since Θ admits selection operators for natural numbers there is a Θ-computable partial function φ such that if $B_a \neq \emptyset$ then

$\varphi(a)$ is an index for the characteristic function of a nonempty subset of B_a. If $\varphi(a)\downarrow$ then $\lambda x\{\varphi(a)\}_\Theta(x,a)$ is the characteristic function of a nonempty subset of B_a. Hence $B_a \neq \emptyset \Longleftrightarrow \varphi(a)\downarrow$. Hence $\exists xP(x,a) \Longleftrightarrow \varphi(a)\downarrow$. This contradicts the fact that $\exists xP(x,a)$ is not Θ-semicomputable. Hence there is an a such that $B_a \neq \emptyset$, and there is no nonempty subset of B_a which is Θ-computable in a. □

<u>Lemma 22</u>: Suppose $x \in B \Longleftrightarrow \{e\}_\Theta(x,a)\downarrow$. Then there is a subset of B which is nonempty and Θ-computable in a iff $\exists x|\{e\}_\Theta(x,a)| < \kappa^a$.

Proof: Suppose B' is a nonempty subset of B, and B' is Θ-computable in a. There is an index e' such that $\{e'\}_\Theta(a)\downarrow$ iff $\forall x(x \in B' \Rightarrow \{e\}_\Theta(x,a)\downarrow)$, and $\{e'\}_\Theta(a)\downarrow \Longrightarrow |\{e'\}_\Theta(a)| > |\{e\}_\Theta(x,a)|$ for all $x \in B'$. Now $\{e'\}_\Theta(a)\downarrow$, and $|\{e'\}_\Theta(a)| < \kappa^a$. Since B' is nonempty $\exists x|\{e\}_\Theta(x,a)| < \kappa^a$.

Suppose $\exists x|\{e\}_\Theta(x,a)| < \kappa^a$. Then there is an e' such that $\{c'\}_\Theta(a)\downarrow$ and $\exists x|\{e\}_\Theta(x,a)| < |\{e'\}_\Theta(a)|$. Let B' be the subset of B defined by: $x \in B' \Longleftrightarrow |\{e\}_\Theta(x,a)| < |\{e'\}_\Theta(a)|$. Then B' is nonempty and Θ-computable in a. □

<u>Lemma 23</u>: For all e,a: If $\exists x|\{e\}_\Theta(x,a)| < \kappa^{S,a}$ then $\exists x|\{e\}_\Theta(x,a)| < \kappa^a$.

<u>Proof:</u> Suppose $\exists x|\{e\}_\Theta(x,a)| < \kappa^{S,a}$. Then $\exists e' \in N\ \exists r \in S[\{e'\}_\Theta(r,a)\downarrow$ and $\exists x|\{e\}_\Theta(x,a)| < |\{e'\}_\Theta(r,a)|]$.

There is an index e'' such that $\{e''\}_\Theta(a)\downarrow$ iff $\exists e' \in N\ \exists r \in S[\{e'\}_\Theta(r,a)\downarrow$ and $\exists x|\{e\}_\Theta(x,a)| < |\{e'\}_\Theta(r,a)|]$,

and if $\{e''\}_\Theta(a)\downarrow$ then $|\{e''\}_\Theta(a)| > |\{e'\}_\Theta(r,a)|$ for some e',r. (e'' exists because Θ admits selection operators for numbers, and S is strongly Θ-finite. The quantifier $\exists x$ can be expressed by E since the relation $\lambda x|\{e\}_\Theta(x,a)| < |\{e'\}_\Theta(r,a)|$ is Θ-computable in a when $\{e'\}_\Theta(r,a)\downarrow$.) Now $\{e''\}_\Theta(a)\downarrow$, and $|\{e''\}_\Theta(a)| < \kappa^a$.

There are e',r such that $|\{e'\}_\Theta(r,a)| < |\{e''\}_\Theta(a)|$, and $\exists x|\{e\}_\Theta(x,a)| < |\{e'\}_\Theta(r,a)|$. Hence $\exists x|\{e\}_\Theta(x,a)| < \kappa^a$. □

Lemma 23 motivates the next definition.

Definition: Let σ be an ordinal $\leq \kappa_\Theta$. Then σ is a-reflecting if for all $e \in N$: $\exists x\,|\{e\}_\Theta(x,a)| < \sigma \Longrightarrow \exists x|\{e\}_\Theta(x,a)| < \kappa^a$.

Remark 1: The a-reflecting ordinals are an initial segment of the ordinals. By lemma 23 $\kappa^{S,a}$ is a-reflecting. In [21] this fact is called "simple reflection". If the Θ-semicomputable relations are not closed under existential quantifiers then by an earlier discussion and lemma 22 κ_Θ is not a-reflecting for all a.

If σ is a limit of a-reflecting ordinals then σ is a-reflecting. Hence there is a greatest a-reflecting ordinal. This ordinal is denoted by κ_r^a.

If $S = Tp(0) \cup \ldots \cup Tp(n-1)$ where $n > 1$, and $(\Theta, ||_\Theta)$ is the normal computation theory obtained from a normal list then it can be proved that κ_Θ is not a-reflecting for any a. $\kappa^a < \kappa^{S,a}$ because $n > 1$. Hence $\kappa^a < \kappa^{S,a} \leq \kappa_r^a < \kappa_\Theta$. In the following pages it will be proved that $\kappa^{S,a} < \kappa_r^a$ for all normal computation theories and all a.

Remark 2: There is an equivalent way of defining the a-reflecting ordinals. For $\tau < \kappa_\Theta$ let $H_\tau = \{\langle e,a\rangle : |\langle e,a\rangle|_\Theta < \tau\}$. Let $\mathcal{F}$ be the class of formulas in the 1.order language which has a symbol for each of the following functions and predicates: N, s, S, M, K, L, $\lambda x i(x)_i$, $\langle\circ\circ\circ\rangle_n$, Seq, lh. The language also has a unary predicate symbol X and a constant symbol for the number 0. $\sigma \leq \kappa_\Theta$ is a-reflecting if for each $\varphi(x,X)$ in $\mathcal{F}$: $\exists\tau < \sigma\ \varphi(a,H_\tau) \Longrightarrow \exists\tau < \kappa^a\ \varphi(a,H_\tau)$. This definition will not be used in this paper.

Lemma 24: Suppose $x \in B \Longleftrightarrow \{e\}_\Theta(x,a)\downarrow$. Then i), ii) and iii) are equivalent.

i) There is a subset B' of B which is nonempty and Θ-computable in a.

ii) $\exists x |\{e\}_{\Theta}(x,a)| < \kappa^{a}$

iii) $\exists x |\{e\}_{\Theta}(x,a)| < \kappa_{r}^{a}$.

The next lemma is a characterization of the subsets of I which are strongly Θ-finite.

<u>Lemma 25</u>: Let $B \subseteq I$ be Θ-computable. Then i), ii), and iii) are equivalent.

i) B is strongly Θ-finite,

ii) for all e,a: $\exists x \in B\{e\}_{\Theta}(x,a)\downarrow \implies \exists x \in B|\{e\}_{\Theta}(x,a)| < \kappa^{a}$,

iii) for all a and all subsets C of B: if C is nonempty and Θ-semicomputable in a then there is a nonempty subset C' of C which is Θ-computable in a.

<u>Proof</u>: i) $\Rightarrow$ ii). Let $\mathcal{F}_B$ be the partial functional defined by:

$$\mathcal{F}_B(\varphi) \simeq \begin{cases} 0 & \text{if } \exists x \in B \quad \varphi(x) \simeq 0 \\ 1 & \text{if } \forall x \in B \quad \exists y \neq 0 \quad \varphi(x) \simeq y \end{cases}$$

By definition B is strongly Θ-finite iff the functional $\mathcal{F}_B$ is weakly Θ-computable. Suppose B is strongly Θ-finite and $\exists x \in B\{e\}_{\Theta}(x,a)\downarrow$. Choose an index $\hat{e}$ such that $\{\hat{e}\}_{\Theta}(x,a) \simeq t(\{e\}_{\Theta}(x,a))$, where t is the constant function with value 0, and $|\{\hat{e}\}_{\Theta}(x,a)| > |\{e\}_{\Theta}(x,a)|$ when $\{e\}_{\Theta}(x,a)\downarrow$. Then $\mathcal{F}_B(\varphi) \simeq 0$ where $\varphi = \lambda x\{\hat{e}\}_{\Theta}(x,a)$. Since $\mathcal{F}_B$ is weakly Θ-computable there is an index e' and a subfunction φ' of φ such that $\mathcal{F}_B(\varphi') \simeq 0$, $\{e'\}_{\Theta}(a) \simeq 0$, $|\{e'\}_{\Theta}(a)| > |\{\hat{e}\}_{\Theta}(x,a)|$ for all $x \in \mathrm{dom}\,\varphi'$. Since $\mathcal{F}_B(\varphi') \simeq 0$ there is an x in B such that $\varphi'(x) \simeq 0$. Hence $\inf\{|\{\hat{e}\}_{\Theta}(x,a)| : x \in B \text{ and } \varphi(x) \simeq 0\} < |\{e'\}_{\Theta}(a)| < \kappa^{a}$.

To prove ii) $\Rightarrow$ i) suppose $\exists x \in B\ \{e\}_{\Theta}(x,a)\downarrow \Rightarrow \exists x \in B|\{e\}_{\Theta}(x,a)| < \kappa^{a}$. As a relation of e, a the relation $\exists x \in B|\{e\}_{\Theta}(x,a)| < \kappa^{a}$ is Θ-semicomputable. For $\exists x \in B|\{e\}_{\Theta}(x,a)| < \kappa^{a} \iff \exists n(\{n\}_{\Theta}(a)\downarrow$ and

$\exists x \in B|\{e\}_\Theta(x,a)| < |\{n\}_\Theta(a)|)$. B is weakly Θ-finite since B is Θ-computable and the computation theory is normal. An index for the relation "$\{n\}_\Theta(a)\downarrow$ and $\exists x \in B|\{e\}_\Theta(x,a)| < |\{n\}_\Theta(a)|$" can be constructed from an index for the functional which proves that B is weakly Θ-finite, and an index for the function which compares lengths of computations. Since Θ admits selection operators for natural numbers, there is a partial Θ-computable function $\varphi(e,a)$ such that if for some n $\{n\}_\Theta(a)\downarrow$ and $\exists x \in B|\{e\}_\Theta(x,a)| < |\{n\}_\Theta(a)|$ then $\varphi(e,a)$ is such an n. Conversely if $\varphi(e,a)\downarrow$ then $\varphi(e,a)$ is such an n. Construct an index e' such that $\{e'\}_\Theta(e,a) \simeq \{\varphi(e,a)\}_\Theta(a)$, and $|\{e'\}_\Theta(e,a)| > |\{\varphi(e,a)\}_\Theta(a)|$ when $\varphi(e,a)\downarrow$. Then $\exists x \in B\{e\}_\Theta(x,a)\downarrow$ iff $\{e'\}_\Theta(e,a)\downarrow$, and if $\exists x \in B\{e\}_\Theta(x,a)\downarrow$ then $\exists x \in B|\{e\}_\Theta(x,a)| < |\{e'\}_\Theta(e,a)|$. Construct e" from e' such that $\exists x \in B\{e\}_\Theta(x,a) \simeq 0$ iff $\{e"\}_\Theta(e,a)\downarrow$, and if $\{e"\}_\Theta(e,a)\downarrow$ then $\exists x \in B(\{e\}_\Theta(x,a) \simeq 0$ and $|\{e\}_\Theta(x,a)| < |\{e"\}_\Theta(e,a)|)$. Since B is weakly Θ-finite there is an index e‴ such that $\forall x \in B\ \exists y \neq 0\ \{e\}_\Theta(x,a) \simeq y$ iff $\{e'''\}_\Theta(e,a) \simeq 1$, and if $\{e'''\}_\Theta(e,a) \simeq 1$ then $|\{e\}_\Theta(x,a)| < |\{e"\}_\Theta(e,a)|$ for all $x \in B$. An index for $\mathcal{F}_B$ can be found from e" and e‴. Hence B is strongly Θ-finite.

ii) $\Rightarrow$ iii). Let C be a nonempty subset of B which is Θ-semicomputable in a. Choose e such that $x \in C \iff \{e\}_\Theta(x,a)\downarrow$. Since C is nonempty $\exists x \in B\{e\}_\Theta(x,a)\downarrow$. By ii) $\exists x \in B|\{e\}_\Theta(x,a)| < \kappa^a$. Choose n such that $\{n\}_\Theta(a)\downarrow$ and $\exists x \in B|\{e\}_\Theta(x,a)| < |\{n\}_\Theta(a)|$. Let $C' = \{x : |\{e\}_\Theta(x,a)| < |\{n\}_\Theta(a)|\}$. Then C' is a nonempty subset of C, and C' is Θ-computable in a.

iii) $\Rightarrow$ ii). Suppose $\exists x \in B\{e\}_\Theta(x,a)\downarrow$. Let $C = \{x : x \in B$ and $\{e\}_\Theta(x,a)\downarrow\}$. Then C is a nonempty subset of B which is Θ-semicomputable in a. By iii) there is a nonempty subset C' of C which is Θ-computable in a. An index e' for the relation "$\forall x(x \in C' \Rightarrow \{e\}_\Theta(x,a)\downarrow)$" can be found such that $\{e'\}_\Theta(a)\downarrow$, and

$|\{e\}_{\Theta}(x,a)| < |\{e'\}_{\Theta}(a)|$ for all $x \in C'$. Hence
$\exists x \in B|\{e\}_{\Theta}(x,a)| < |\{e'\}_{\Theta}(a)| < \varkappa^{a}$. This proves ii). □

Let a be a fixed finite list of individuals. Let
$P = \{\langle e,b^{-}\rangle : \{e\}_{\Theta}(b)\downarrow$ and b is a list of elements from S and the list a. b^{-} is the list obtained by removing the elements of a from b.$\}$

P is a subset of S, hence an element of ${}^{S}\omega$.
P is a complete Θ-semicomputable in a, subset of S, i.e. all other subsets of S which are Θ-semicomputable in a can be reduced to P by a Θ-computable one-one mapping.

<u>Theorem 8</u>: $\varkappa^{S,P,a}$ is a-reflecting.
($\varkappa^{S,P,a} = \varkappa^{X}$ where $X = S \cup \{P\} \cup \{x: x$ is in the list $a\}$.)
In [21] this result is called "further reflection". To prove theorem 8 we first prove three propositions.

Let pwo_S denote the set of prewellorderings with domain $\subseteq S$.
If $pwo_S(x)$ let $|x|$ denote the length of the prewellordering.
If $y \in dom(x)$ let $|y|_x$ denote the length of that part of the prewellordering which is below y. There is a natural prewellordering with domain P and length $\lambda^{S,a}$ defined by: $(\langle e,b^{-}\rangle,\langle e',b'^{-}\rangle)$ is in the prewellordering if $|\{e\}_{\Theta}(b)|_{\Theta} < |\{e'\}_{\Theta}(b')|_{\Theta}$ for $\langle e,b^{-}\rangle,\langle e',b'^{-}\rangle \in P$. If x is in this prewellordering let $|x|_P$ denote the length of that part of the prewellordering which is below x.

<u>Proposition 1</u>: The relation "$pwo_S(x)$ and $|x| < \lambda^{S,a}$" is Θ-semicomputable as a relation of x,a. If $|x| \geq \lambda^{S,a}$ then P is Θ-computable in x,a and possibly an element in S.

Proof: The relation $pwo_S(x)$ is Θ-computable, for $pwo_S(x) \Longleftrightarrow$ x is a linear preordering, and there are no infinite descending chains in the ordering. Since the domain of the ordering is a subset of S an infinite descending chain in the ordering can be regarded as an element of ${}^{S}\omega$. Since ${}^{S}\omega$ is weakly finite the expression "there are no infinite descending chains" is Θ-computable.

There is a Θ-computable partial function φ such that $\varphi(x,y,z,a) \simeq 0$ iff $\mathrm{pwo}_S(x)$ and $z \in \mathrm{dom}\,x$, $y \in P$, $|z|_x \leq |y|_P$. (To prove that such a φ exists use the first recursion theorem.) Then $\mathrm{pwo}_S(x)$ and $|x| < \lambda^{S,a} \iff \mathrm{pwo}_S(x)$ and $\exists y[y \in P$ and $\forall z \in \mathrm{dom}\,x\ \varphi(x,y,z,a) \simeq 0]$.
Since the quantifier $\exists y$ ranges over S the right hand side of the equivalence is Θ-semicomputable.

By another application of the first recursion theorem one can prove that there is a Θ-computable partial function ψ such that

$$\psi(x,y,z,a) \simeq 0 \text{ if } \mathrm{pwo}_S(x),\ z \in \mathrm{dom}\ x,\ |z|_x \leq |y|_P$$
$$\simeq 1 \text{ if } \mathrm{pwo}_S(x),\ z \in \mathrm{dom}\ x,\ y \in P \text{ and } |z|_x > |y|_P$$

(Convention: $|y|_P = \lambda^{S,a}$ if $y \notin P$.)

Suppose $|x| \geq \lambda^{S,a}$. In a simple way one can define a prewellordering x' from x such that $|x'| = |x| + 1$. Then $|x'| > \lambda^{S,a}$. There is an r in $\mathrm{dom}(x')$ such that $|r|_{x'} = \lambda^{S,a}$. $\lambda y(1 - \psi(x',y,r,a))$ is the characteristic function of P. Hence P is Θ-computable in x,a,r. □ (prop. 1)

<u>Proposition 2</u>: If P is Θ-computable in a,x and elements from S then $\kappa^{S,P,a} \leq \kappa^{S,x,a}$.

Proof: Let $\nu < \kappa^{S,P,a}$ be arbitrary. Then for some e,b: $\{e\}_\Theta(b)\downarrow$ and $\nu < |\{e\}_\Theta(b)|$, where the list b can contain P and elements from a and S. Since P is Θ-computable in a,x and elements $r_1 \ldots r_k \in S$ there is an index e' such that $\lambda y\{e'\}_\Theta(y,x,a,r_1 \ldots r_k)$ is the characteristic function of P. Substitute this function for P in $\{e\}_\Theta(b)$ and obtain a computation $\{e''\}_\Theta(c)$ such that $\{e''\}_\Theta(c)\downarrow$ and $|\{e\}_\Theta(b)| < |\{e''\}_\Theta(c)|$, where the list c can contain x and elements from a and S. $|\{e''\}_{\Theta}(c)| < \varkappa^{S,x,a}$. Hence $\nu < \varkappa^{S,x,a}$, and $\varkappa^{S,P,a} \leq \varkappa^{S,x,a}$. $\square$ (prop. 2)

Proposition 3: There is an index $\hat{e}$ such that if $\exists y|\{e\}_{\Theta}(y,a)| < \varkappa^{S,a,x}$ then $\{\hat{e}\}_{\Theta}(e,a,x)\downarrow$ and $\exists y|\{e\}_{\Theta}(y,a)| < |\{\hat{e}\}_{\Theta}(e,a,x)|$.

Proof: $\exists y|\{e\}_{\Theta}(y,a)| < \varkappa^{S,a,x} \iff \exists e' \exists r \in S[\{e'\}_{\Theta}(r,x,a)\downarrow$ and $\exists y|\{e\}_{\Theta}(y,a)| \leq |\{e'\}_{\Theta}(r,x,a)|]$. Let $C = \{\langle e',r\rangle: \{e'\}_{\Theta}(r,x,a)\downarrow$ and $\exists y|\{e\}_{\Theta}(y,a)| \leq |\{e'\}_{\Theta}(r,x,a)|\}$. Then C is a subset of S which is Θ-semicomputable in a,x. Suppose $\exists y|\{e\}_{\Theta}(y,a)| < \varkappa^{S,a,x}$. Then C is nonempty. S is strongly Θ-finite since $(\Theta, ||_{\Theta})$ is normal. By lemma 25 there is a nonempty subset C' of C which is Θ-computable in a,x. An index for C' can be found uniformly from an index for C. There is an index $\hat{e}$ for the relation "$\forall e',r(\langle e',r\rangle \in C' \Longrightarrow \{e'\}_{\Theta}(r,x,a)\downarrow)$" such that $\{\hat{e}\}_{\Theta}(e,a,x)\downarrow$ and $|\{e'\}_{\Theta}(r,x,a)| < |\{\hat{e}\}_{\Theta}(e,a,x)|$ for all $\langle e',r\rangle \in C'$. Since $\exists y|\{e\}_{\Theta}(y,a)| < |\{e'\}_{\Theta}(r,x,a)|$ for all $\langle e',r\rangle \in C'$, $\exists y|\{e\}_{\Theta}(y,a)| < |\{\hat{e}\}_{\Theta}(e,a,x)|$. $\square$ (prop. 3)

Proof of theorem 8: Suppose $\exists y|\{e\}_{\Theta}(y,a)| < \varkappa^{S,P,a}$. If $pwo_S(x)$ then either $|x| < \lambda^{S,a}$ or $|x| \geq \lambda^{S,a}$. In the latter case P is Θ-computable in a,x and elements from S by proposition 1. By proposition 2 $\varkappa^{S,P,a} \leq \varkappa^{S,x,a}$, hence $\exists y|\{e\}_{\Theta}(y,a)| < \varkappa^{S,x,a}$. Thus for all x: $pwo_S(x) \Longrightarrow |x| < \lambda^{S,a}$ or $\exists y|\{e\}_{\Theta}(y,a)| < \varkappa^{S,x,a}$. By proposition 1 the relation "$pwo_S(x)$ and $|x| < \lambda^{S,a}$" is Θ-semicomputable. Let e_1 be an index for this relation. By proposition 3 there is an index $\hat{e}$ such that if $\exists y|\{e\}_{\Theta}(y,a)| < \varkappa^{S,x,a}$ then $\{\hat{e}\}_{\Theta}(e,a,x)\downarrow$ and $\exists y|\{e\}_{\Theta}(y,a)| < |\{\hat{e}\}_{\Theta}(e,a,x)|$. There is an index f such that $\{f\}_{\Theta}(e,a)\downarrow$ if for all x: $pwo_S(x) \Longrightarrow |x| < \lambda^{S,a}$ or $\exists y|\{e\}_{\Theta}(y,a)| < \varkappa^{S,x,a}$, and in this case $|\{f\}_{\Theta}(e,a)| > \inf\{|\{e_1\}_{\Theta}(x,a)|, |\{\hat{e}\}_{\Theta}(e,a,x)|\}$ for all x such that $pwo_S(x)$. Now $\{f\}_{\Theta}(e,a)\downarrow$.

Choose x such that $pwo_S(x)$ and $|x| > \lambda^{S,a}$. Then $\{e_1\}_\Theta(x,a)\uparrow$, and $\exists y|\{e\}_\Theta(y,a)| < \kappa^{S,x,a}$. Hence $\exists y|\{e\}_\Theta(y,a)| < |\{\hat{e}\}_\Theta(e,a,x)| < |\{f\}_\Theta(e,a)| < \kappa^a$. □

<u>Remark</u>: $\kappa^{S,a} < \kappa^{P,a} \leq \kappa^{S,P,a}$. To prove that $\kappa^{S,a} < \kappa^{P,a}$ let e be an index for the expression $\forall\langle e',b^-\rangle\ (\langle e',b^-\rangle \in P \Longrightarrow \{e'\}_\Theta(b)\downarrow)$ such that $\{e\}_\Theta(P,a)\downarrow$ and $|\{e'\}_\Theta(b)| < |\{e\}_\Theta(P,a)|$ for all $\langle e',b^-\rangle$ in P. Since $\kappa^{S,a} = \sup\{|\{e'\}_\Theta(b)| : \langle e',b^-\rangle \in P\}$, $\kappa^{S,a} \leq |\{e\}_\Theta(P,a)|$. Hence $\kappa^{S,a} < \kappa^{P,a}$ since $|\{e\}_\Theta(P,a)| < \kappa^{P,a}$.

<u>Corollary</u>: Suppose that B is a set of subsets of S such that B is Θ-semicomputable in a, and B contains a subset of S which is nonempty and Θ-semicomputable in a. Then B contains a subset of S which is nonempty and Θ-computable in a.

Proof: Let $x \in B \Longleftrightarrow \{e\}_\Theta(x,a)\downarrow$. Let $C \in B$ be Θ-semicomputable in a with index e', i.e. $r \in C \Longleftrightarrow \{e'\}_\Theta(r,a)\downarrow$. For $\sigma < \kappa_\Theta$ let C_σ be defined by: $C_\sigma = \{r : |\{e'\}_\Theta(r,a)| < \sigma\}$. If $\sigma \geq \kappa^{S,a}$ then $C_\sigma = C$. If y is a convergent computation and $|y|_\Theta = \sigma$ then C_σ is Θ-computable in a,y, uniformly. Let g be an index such that $\{g\}_\Theta(y,a)\downarrow$ iff $C_{|y|_\Theta} \in B$. By the remark $\kappa^{S,a} < \kappa^{P,a}$. Choose $y = \langle f,P,a\rangle$ such that $\{f\}_\Theta(P,a)\downarrow$ and $\kappa^{S,a} \leq |\{f\}_\Theta(P,a)|$. Then $C_{|y|_\Theta} = C$, $C_{|y|_\Theta} \in B$, $\{g\}_\Theta(y,a)\downarrow$, and $|\{g\}_\Theta(y,a)| < \kappa^{P,a}$. By theorem 8 $\kappa^{S,P,a}$ is reflecting. Hence $\kappa^{P,a}$ is reflecting, and $\exists y|\{g\}_\Theta(y,a)| < \kappa^a$. Choose $h \in \omega$ such that $\{h\}_\Theta(a)\downarrow$ and $\exists y|\{g\}_\Theta(y,a)| < |\{h\}_\Theta(a)|$. The set $D = \{y : |\{g\}_\Theta(y,a)| < |\{h\}_\Theta(a)|$ and $|y|_\Theta$ is minimal$\}$ is Θ-computable in a. Let $\tau = |y|_\Theta$ where $y \in D$. Then $C_\tau \in B$ because $\{g\}_\Theta(y,a)\downarrow$ when $y \in D$.
$r \in C_\tau \Longleftrightarrow \forall y(y \in D \Rightarrow |\{e'\}_\Theta(r,a)| < |y|_\Theta)$. The relation inside the brackets is Θ-computable in a. Hence C_τ is Θ-computable in a. □

Notations:

$sc(\Theta) = \{X : X \subseteq I$ and X is Θ-computable$\}$

$sc(\Theta,a) = \{X : X \subseteq I$ and X is Θ-computable in $a\}$

$S\text{-}sc(\Theta,a) = \{X : X \subseteq S$ and X is Θ-computable in $a\}$

$en(\Theta) = \{X : X \subseteq I$ and X is Θ-semicomputable$\}$

$S\text{-}en(\Theta) = \{X : X \subseteq S$ and X is Θ-semicomputable$\}$

If $\mathcal{L}$ is a normal list we write $en(\mathcal{L})$, $sc(\mathcal{L})$ instead of $en(\Theta)$, $sc(\Theta)$, where $(\Theta, ||_\Theta)$ is the normal computation theory obtained from $\mathcal{L}$.

If $F \in Tp(m)$ then for $n > 0$:

$n - sc(F) = \{X : X \subseteq Tp(n-1)$ and X is recursive in $F\}$

$n - en(F) = \{X : X \subseteq Tp(n-1)$ and X is recursively enumerable in $F\}$

In the rest of this chapter the following problem will be considered. Let $(\Theta, ||_\Theta)$ be a normal computation theory. Is there a list $\mathcal{L}$ such that $(\Theta, ||_\Theta)$ is similar to recursion in $\mathcal{L}$? By "$(\Theta, ||_\Theta)$ is similar to recursion in $\mathcal{L}$" we will mean one of the following two statements: "$en(\Theta) = en(\mathcal{L})$"; "for all partial functions φ: φ is Θ-computable iff φ is recursive in $\mathcal{L}$". (When $S = \omega$ these two statements are equivalent.)

We have the following result from recursion in higher types: If F is a normal object of type $> n+2$ then there is no normal object G of type $n+2$ such that $n+1 - en(F) = n+1 - en(G)$. The case $n = 0$ is proved by Moschovakis in [17]. When $n > 0$ then $n+1 - en(F)$ is closed under the quantifier $\exists x \in Tp(n)$, and $n+1 - en(G)$ is not for any $G \in Tp(n+2)$. Hence $n+1 - en(F) \neq n+1 - en(G)$.

This result can be translated to the setting of this paper. Let F be as above. Let $S = Tp(0) \cup \ldots \cup Tp(n-1)$, let $\mathfrak{A}$ be the computation domain with subindividuals S, and let $(\Theta, ||_\Theta)$ be the normal computation theory on $\mathfrak{A}$ obtained from F as described in § 3.

Then there is no normal list $\mathcal{L}$ such that any of the two statements above is true.

The following two results are proved for recursion on the types: If F is a normal object of type $> n+2$ then there is a normal object G of type $n+2$ such that $n{+}1-\mathrm{sc}(F) = n{+}1-\mathrm{sc}(G)$. If F is a normal object to type $> n+2$ where $n > 0$ then there is a normal object G of type $n+2$ such that $n-\mathrm{en}(F) = n-\mathrm{en}(G)$. The first result is called the "+1 theorem" and is proved by Sacks in [20] and [21]. The other result is called the "+2 theorem" and is proved by Harrington in [7].

Theorem 9 is a generalization of these two theorems. Both theorems are corollaries of theorem 9 when $n > 0$: If F is a normal object of type $> n+2$ where $n > 0$, let $(\Theta, ||_\Theta)$ be the normal computation theory obtained from F as mentioned above. By theorem 9 there is a normal list $\mathcal{L}$ such that $S-\mathrm{en}(\Theta) = S-\mathrm{en}(\mathcal{L})$, $\mathrm{sc}(\Theta,r) = \mathrm{sc}(\Theta,r)$ for all $r \in S$. Hence there is a normal object G of type $n+2$ such that $n-\mathrm{en}(F) = n-\mathrm{en}(G)$ and $n{+}1-\mathrm{sc}(F,x) = n{+}1-\mathrm{sc}(G,x)$ for all $x \in Tp(n-1)$.

<u>Theorem 9</u>: Let $(\Theta, ||_\Theta)$ be normal. Then there is a normal list $\mathcal{L}$ such that $S-\mathrm{en}(\Theta) = S-\mathrm{en}(\mathcal{L})$, and $\mathrm{sc}(\Theta,r) = \mathrm{sc}(\mathcal{L},r)$ for all $r \in S$.

Proof: Let $P = \{\langle e,a\rangle : \{e\}_\Theta(a)\downarrow$ and a is a list of objects from $S\}$. An ordinal η is <u>Θ-subconstructive</u> if $\eta = |\{e\}_\Theta(a)|_\Theta$ for some $\langle e,a\rangle$ in P. The order type of the Θ-subconstructive ordinals is $\lambda = \lambda^S$. For $\nu < \lambda$ let η_ν be the ν-th Θ-subconstructive ordinal.

The list $\mathcal{L}$ will contain three objects: the functionals E and G and the equality relation on S. G is constructed in stages. If τ is an ordinal then G_τ is a partial functional. The domain of G_τ ($\mathrm{dom}\, G_\tau$) contains that part of $\mathrm{dom}\, G$ which is needed to generate H_τ, where $H_\tau = \{\langle e,a\rangle : |\{e\}^{\mathcal{L}}(a)| < \tau\}$. If $\tau < \tau'$ then $G_\tau \subseteq G_{\tau'}$, i.e. if $f \in \operatorname{dom} G_\tau$ then $f \in \operatorname{dom} G_{\tau'}$, and $G_\tau(f) = G_{\tau'}(f)$. Finally G is defined by:

$$G(f) = \begin{cases} (\bigcup_\tau G_\tau)(f) & \text{if } f \in \bigcup_\tau \operatorname{dom} G_\tau \\ 0 & \text{otherwise} \end{cases}$$

Let σ_ν be the ν-th $\mathcal{L}$-subconstructive ordinal.

Suppose G_τ, H_τ are defined.

$H_{\tau+1} = \{\langle e,a\rangle :$ all immediate subcomputations of $\{e\}^{\mathcal{L}}(a)$ are in $H_\tau\}$. If $\forall x(\{e\}^{\mathcal{L}}(x,a)\downarrow$ and $|\{e\}^{\mathcal{L}}(x,a)| < \tau)$ then $\langle\langle 16,n,e,2\rangle,a\rangle \in H_{\tau+1}$ $(\{\langle 16,n,e,2\rangle\}^{\mathcal{L}}(a) \simeq G(\lambda x\{e\}^{\mathcal{L}}(x,a)))$. Let $f = \lambda x\{e\}^{\mathcal{L}}(x,a)$. If f is not already in the domain of G_τ we must define $G(f)$ at this stage. Hence we let $f \in \operatorname{dom} G_{\tau+1}$, and $G_{\tau+1}(f) = 0$ or 1. $G_{\tau+1}$ is said to be the trivial extension of G_τ if $\operatorname{dom} G_{\tau+1} = \operatorname{dom} G_\tau \cup \{f : f$ is as above$\}$, and $G_{\tau+1}(f) = 0$ when $f \in \operatorname{dom} G_{\tau+1} - \operatorname{dom} G_\tau$. In the construction of $G_{\tau+1}$ we let $G_{\tau+1}(f) = 0$ when f is as above and $f \notin \operatorname{dom} G_\tau$, if not otherwise mentioned.

To obtain $S - en(\Theta) = S - en(\mathcal{L})$ information about Θ must be brought into the construction of G. This is done at stages which are $\mathcal{L}$-subconstructive. Suppose we have constructed G_μ for all $\mu < \tau$, and $\tau = \sigma_\nu$ for a $\nu < \lambda$. Regard the set $\{x : |x|_\Theta \leq \eta_\nu\}$. For each x in this set we take a function f $(= f_{xy})$ such that $f \notin \operatorname{dom} G_\mu$ when $\mu < \tau$, and let $G_\tau(f) = 1$. From f and G_τ one can regain the information that x is in the above set. f is also recursive in $\mathcal{L},x,y$, where y is any $\mathcal{L}$-computation of length τ. The existence of such a set of functions is proved in the following proposition.

<u>Proposition 1</u>: Let $\mathcal{L}$ be any normal list. Let y be a convergent $\mathcal{L}$-computation with length τ. For each $x \in I$ there is a total function f_{xy} such that f_{xy} is recursive in $\mathcal{L},x,y$, and if

$x \neq x'$ then $f_{xy} \neq f_{x'y}$. If $f_{xy} = \lambda t\{e\}^{\mathcal{L}}(t,a)$ for some e,a then $\tau \leq |\{e\}^{\mathcal{L}}(t,a)|$ for some $t \in I$.

Proof: Let τ^+ be the least limit ordinal $\geq \tau$. The set of $\mathcal{L}$-computations with length $< \tau^+$ is recursive in $\mathcal{L},y$. Let f_y be the function defined by: $f_y(u) = 0$ if u is not a sequence $\langle e,a\rangle$. If $u = \langle e,a\rangle$ then see if $|\{e\}^{\mathcal{L}}(t,a)| < \tau^+$ for all t. If not let $f_y(u) = 0$. If true let $f_y(u)$ be something different from $\{e\}^{\mathcal{L}}(\langle e,a\rangle,a)$. Let for instance $f_y(\langle e,a\rangle) = 1$ if $\{e\}^{\mathcal{L}}(\langle e,a\rangle,a) \simeq 0$, $f_y(\langle e,a\rangle) = 0$ otherwise. f_y is recursive in $\mathcal{L},y$. If $f = \lambda t\{e\}^{\mathcal{L}}(t,a)$ is a total function such that $|\{e\}^{\mathcal{L}}(t,a)| < \tau^+$ for all t then f_y and f have different values for $t = \langle e,a\rangle$. Hence $f \neq f_y$. Let

$$\begin{aligned} f_{xy}(t) &= \langle f_y(t),x,0\rangle && \text{if } x \in S \\ &= \langle f_y(t),x(t),1\rangle && \text{if } x \in {}^S\omega,\ t \in S \\ &= f_y(t) && \text{if } x \in {}^S\omega,\ t \in {}^S\omega \end{aligned}$$

Then f_{xy} is recursive in $\mathcal{L},x,y$, uniformly in x,y. If $x \neq x'$ then $f_{xy} \neq f_{x'y}$. If $f_{xy} = \lambda t\{e\}^{\mathcal{L}}(t,a)$ then $\tau \leq |\{e\}^{\mathcal{L}}(t,a)|$ for some t. For suppose $|\{e\}^{\mathcal{L}}(t,a)| < \tau$ for all t. The function f_y can easily be regained from f_{xy}. In fact there is an index e' such that $f_y = \lambda t\{e'\}^{\mathcal{L}}(t,a)$ and $|\{e'\}^{\mathcal{L}}(t,a)| < |\{e\}^{\mathcal{L}}(t,a)| + \omega$ for all t. Hence $|\{e'\}^{\mathcal{L}}(t,a)| < \tau^+$ for all t. Hence $f_y(\langle e',a\rangle) \neq \{e'\}^{\mathcal{L}}(\langle e',a\rangle,a)$, a contradiction. □ (prop. 1)

Construction of G_τ:

Suppose G_μ, H_μ are defined for all $\mu < \tau$. We define G_τ in two cases:

Case 1: <u>There is an ordinal $\nu < \lambda$ such that ν is the order type of the ordinals $< \tau$ which are $\mathcal{L}$-subconstructive</u>

This case is divided into two subcases:

I : τ is $\mathcal{L}$-subconstructive (i.e. $\tau = \sigma_\nu$) .

Let $G_\tau(f_{xy}) = 1$ for all x,y such that x is a $\circledast$-computation of length $\leq \eta_\nu$, y is an $\mathcal{L}$-computation of length τ.

II : τ is not $\mathcal{L}$-subconstructive (i.e. $\tau < \sigma_\nu$).

Let $\varepsilon = \eta_\nu - \sup\{\eta_\rho : \rho < \nu\}$. The definition of G_τ depends on the answer to the following question: Is there an ordinal π such that $\tau < \pi \leq \tau + \varepsilon$ and π is $\mathcal{L}^0$-subconstructive?

Yes: $G_\tau = \bigcup_{\mu<\tau} G_\mu$ if $\lim \tau$, G_τ is the trivial extension of G_μ if $\tau = \mu + 1$.

No : $G_\tau(f_{xy}) = 1$ when x,y are as in I.

Case 2 : Otherwise

$G_\tau = \bigcup_{\mu<\tau} G_\mu$ if $\lim \tau$, G_τ is the trivial extension of G_μ if $\tau = \mu + 1$.

Remark: One can decide whether or not τ is $\mathcal{L}$-subconstructive before G_τ is defined. For H_μ is defined when $\mu < \tau$. If τ is a limit ordinal then τ is $\mathcal{L}$-subconstructive iff there are $e \in N$ and a list a of objects from S such that $\{e\}^{\mathcal{L}}(t,a)\downarrow$ for all t, and $\sup\{|\{e\}^{\mathcal{L}}(t,a)| + 1 : t \in I\} = \tau$ iff there are e and a as above such that $\langle e,t,a\rangle \in \bigcup_{\mu<\tau} H_\mu$ for all t, and if $\mu < \tau$ then for some t $\langle e,t,a\rangle \notin H_\mu$. If τ is a successor ordinal there is a similar test to decide from $\bigcup_{\mu<\tau} H_\mu$ whether or not τ is $\mathcal{L}$-subconstructive.

The list $\mathcal{L}^0$ in subcase II of case 1 consists of the two functionals G^0, E and the equality relation on S. Hence $\mathcal{L}^0$ is normal. G^0 is defined by:

$G^0(f) = (\bigcup_{\mu<\tau} G_\mu)(f)$ if $\exists\mu < \tau$ $f \in \operatorname{dom} G_\mu$,

$G^0(f) = 0$ otherwise.

The functions f_{xy} are as in proposition 1 with $\mathcal{L} = \mathcal{L}^o$. Then f_{xy} is recursive in $\mathcal{L}^o,x,y$, hence in $\mathcal{L},x,y$ since G^o is recursive in $\mathcal{L},y$. Also $f_{xy} \notin \operatorname{dom} G_\mu$ when $\mu < \tau$.

This defines G, and we prove that $\mathcal{L}$ has the desired properties.

Proposition 2: The order type of the $\mathcal{L}$-subconstrucive ordinals is at least λ.

Proof: Suppose not. Let $\nu < \lambda$ be the order type of the $\mathcal{L}$-subconstructive ordinals. Let $\tau = \sup\{\sigma_\rho : \rho < \nu\}$. τ is the supremum of the $\mathcal{L}$-subconstructive ordinals, and τ is not $\mathcal{L}$-subconstructive. When we define G_τ we are in case 1, subcase II. The answer to the question is no. Let $x \in C_\Theta$ (= the set of convergent Θ-computations), $|x|_\Theta = \eta_\nu$, $x \in S$. Then $G(f_{xy}) = 1$ for all y such that $|y| = \tau$. (Such a y exists because there are $\mathcal{L}$-computations with length greater than all $\mathcal{L}$-subconstructive ordinals.) Also τ is the least ordinal such that $\exists y[|y| = \tau$ and $G(f_{xy}) = 1]$.

There are indexes e_1, e_2 such that $G(f_{xy}) = 1$ iff $\{e_1\}^{\mathcal{L}}(x,y)\downarrow$, and if $G(f_{xy}) = 1$ then $|y|^{\mathcal{L}} \leq |\{e_1\}^{\mathcal{L}}(x,y)|$. For all x,y: $G(f_{xy}) \simeq \{e_2\}^{\mathcal{L}}(x,y)$, and if $G(f_{xy}) = 1$ then $|\{e_1\}^{\mathcal{L}}(x,y)| \leq |\{e_2\}^{\mathcal{L}}(x,y)|$.

Let $Q = \{\langle e,a\rangle : \{e\}^{\mathcal{L}}(a)\downarrow,\ a$ is a list of objects from $S\}$. By theorem 8 the ordinal $\kappa^{Q,x}$ is x-reflecting. By the remark following theorem 8 $\kappa^{S,x} < \kappa^{Q,x}$. $\kappa^{S,x} = \kappa^S$ since $x \in S$. By assumption $\tau = \kappa^S$.

Let m be an index such that $\{m\}^{\mathcal{L}}(Q,x)\downarrow$ and $\kappa^S = \kappa^{S,x} < |\{m\}^{\mathcal{L}}(Q,x)|$. (Such an m exists since $\kappa^{S,x} < \kappa^{Q,x}$.) Then $\exists y[|y|^{\mathcal{L}} < |\{m\}^{\mathcal{L}}(Q,x)|$ and $G(f_{xy}) = 1]$ (let $|y|^{\mathcal{L}} = \tau$). This expression defines a relation $R(Q,x)$, and R is recursively enumerable in $\mathcal{L}$. An index for the expression inside the brackets can be found by the instructions: First see if $|y|^{\mathcal{L}} < |\{m\}^{\mathcal{L}}(Q,x)|$. If false then stop. If true then compute $\{e_2\}^{\mathcal{L}}(x,y)$. For Q and x such that $\{m\}^{\mathcal{L}}(Q,x)\downarrow$ this defines a total relation of y. Hence the quantifier $\exists y$ can be expressed by E. An index e' for R can be found such that $\{e'\}^{\mathcal{L}}(Q,x)\downarrow$, and if $|y|^{\mathcal{L}} < |\{m\}^{\mathcal{L}}(Q,x)|$ and $G(f_{xy}) = 1$ (for instance when $|y|^{\mathcal{L}} = \tau$) then $|\{e_1\}^{\mathcal{L}}(x,y)| < |\{e'\}^{\mathcal{L}}(Q,x)|$.

Now $|\{e'\}^{\mathcal{L}}(Q,x)| < \kappa^{Q,x}$. Hence $\exists y|\{e_1\}^{\mathcal{L}}(x,y)| < \kappa^{Q,x}$ (let $|y|^{\mathcal{L}} = \tau$). By reflection $\exists y|\{e_1\}^{\mathcal{L}}(x,y)| < \kappa^{x}$. $\kappa^{x} \leq \kappa^{S} = \tau$, and $|y|^{\mathcal{L}} \leq |\{e_1\}^{\mathcal{L}}(x,y)|$. Hence $\exists y[|y|^{\mathcal{L}} < \tau$ and $\{e_1\}^{\mathcal{L}}(x,y)\downarrow]$, i.e. $\exists y[|y|^{\mathcal{L}} < \tau$ and $G(f_{xy}) = 1]$. This is contrary to the fact that τ is the least ordinal such that $\exists y[|y|^{\mathcal{L}} = \tau$ and $G(f_{xy}) = 1]$.

□ (prop. 2)

<u>Proposition 3</u>: $S\text{-en}(\Theta) \subseteq S\text{-en}(\mathcal{L})$; $sc(\Theta,r) \subseteq sc(\mathcal{L},r)$ for all $r \in S$.

Proof: Let $H_\Theta = \{\langle e,a\rangle : |\{e\}_\Theta(a)|_\Theta < \eta_\nu$ for some $\nu < \lambda\}$. Suppose $X \subseteq S$, $X \in S\text{-en}(\Theta)$. Then there is an index e such that $r \in X \Longleftrightarrow \{e\}_\Theta(r)\downarrow$. Hence $r \in X \Longleftrightarrow \langle e,r\rangle \in H_\Theta$.

Suppose $X \subseteq I$, $X \in sc(\Theta,r)$, $r \in S$. Then there are indexes e_1, e_2 such that $x \in X \Longleftrightarrow \{e_1\}_\Theta(x,r)\downarrow$, $x \notin X \Longleftrightarrow \{e_2\}_\Theta(x,r)\downarrow$. There is an index e' such that $\lambda x\{e'\}_\Theta(x,r)$ is the characteristic function of X, and for all $x \in X$: $|\{e_1\}_\Theta(x,r)| < |\{e'\}_\Theta(x,r)|$; for all $x \notin X$: $|\{e_2\}_\Theta(x,r)| < |\{e'\}_\Theta(x,r)|$. Let e be an index for the computation $E(\lambda x\{e'\}_\Theta(x,r))$. Then $\{e\}_\Theta(r)\downarrow$, and for all $x \in X$: $|\{e_1\}_\Theta(x,r)| < |\{e\}_\Theta(r)|$, for all $x \notin X$: $|\{e_2\}_\Theta(x,r)| < |\{e\}_\Theta(r)|$. Now $|\{e\}_\Theta(r)|_\Theta = \eta_\nu$ for some $\nu < \lambda$. Hence $x \in X \Longleftrightarrow \langle e_1,x,r\rangle \in H_\Theta$, $x \notin X \Longleftrightarrow \langle e_2,x,r\rangle \in H_\Theta$.

It is enough to prove that $H_\Theta \in en(\mathcal{L})$, for it follows from the discussion above that if this is true then $S\text{-en}(\Theta) \subseteq S\text{-en}(\mathcal{L})$, and $sc(\Theta,r) \subseteq sc(\mathcal{L},r)$ for all $r \in S$. By proposition 2 $x \in H_\Theta \Longleftrightarrow \exists y \in S\ (y \in C^{\mathcal{L}}$ and $G(f_{xy}) = 1)$. Hence $H_\Theta \in en(\mathcal{L})$.

□ (prop. 3)

Let $\eta = \sup\{\eta_\nu : \nu < \lambda\}$ $(= \kappa^S_\Theta)$, and let $\sigma = \sup\{\sigma_\nu : \nu < \lambda\}$.

Proposition 4: a) There are a total Θ-computable function f and a Θ-computable partial function p such that $\{f(e)\}_\Theta(a) \simeq \{e\}^{\mathcal{L}}(a)$ for all e,a such that $|\{e\}^{\mathcal{L}}(a)| < \sigma$. If $|x|^{\mathcal{L}} < \sigma$ or $|y|^{\mathcal{L}} < \sigma$ then $p(x,y)\downarrow$. If $x \in C^{\mathcal{L}}$, $|x|^{\mathcal{L}} < \sigma$ and $|x|^{\mathcal{L}} \leq |y|^{\mathcal{L}}$, then $p(x,y) \simeq 0$. If $|y|^{\mathcal{L}} < \sigma$ and $|x|^{\mathcal{L}} > |y|^{\mathcal{L}}$ then $p(x,y) \simeq 1$.

b) There are a total Θ-computable function f' and a Θ-computable partial function p' such that $\{f'(e)\}_\Theta(a,P) \simeq \{e\}^{\mathcal{L}}(a)$ for all e,a. If $|x|^{\mathcal{L}} < \kappa^{\mathcal{L}}$ or $|y|^{\mathcal{L}} < \kappa^{\mathcal{L}}$ then $p'(x,y,P)\downarrow$, $p'(x,y,P) \simeq 0$ if $x \in C^{\mathcal{L}}$ and $|x|^{\mathcal{L}} \leq |y|^{\mathcal{L}}$. $p'(x,y,P) \simeq 1$ if $|x|^{\mathcal{L}} > |y|^{\mathcal{L}}$.

Proof: a) Θ-indexes for f and p can be found by the second recursion theorem. We give the main points in the construction of these indexes. Let $\mu < \sigma$. Suppose that $\{f(e)\}_\Theta(a) \simeq \{e\}^{\mathcal{L}}(a)$ for all e,a such that $|\{e\}^{\mathcal{L}}(a)|^{\mathcal{L}} < \mu$, and that $p(x,y)$ is defined and has the right value when $\inf(|x|^{\mathcal{L}}, |y|^{\mathcal{L}}) < \mu$. When $|\{e\}^{\mathcal{L}}(a)|^{\mathcal{L}} = \mu$ we describe $\{f(e)\}_\Theta(a)$ in terms of $\{f(e')\}_\Theta(a')$ and $p(x',y')$, where $\{e'\}^{\mathcal{L}}(a')$ is an immediate subcomputation of $\{e\}^{\mathcal{L}}(a)$ and $\inf(|x'|^{\mathcal{L}}, |y'|^{\mathcal{L}}) < \mu$. When $\inf(|x|^{\mathcal{L}}, |y|^{\mathcal{L}}) = \mu$ we also describe $p(x,y)$ in terms of $p(x',y')$, $\{f(e')\}_\Theta(a')$, where $\inf(|x'|^{\mathcal{L}}, |y'|^{\mathcal{L}}) < \mu$ and $|\{e'\}^{\mathcal{L}}(a')|^{\mathcal{L}} < \mu$.

Let $|\{e\}^{\mathcal{L}}(a)|^{\mathcal{L}} = \mu$. If $\{e\}^{\mathcal{L}}(a)$ is not an application of G it is obvious how to define $f(e)$. So suppose $\{e\}^{\mathcal{L}}(a) \simeq G(\lambda u\{e'\}^{\mathcal{L}}(u,a))$, where $|\{e'\}^{\mathcal{L}}(u,a)| < \mu$ for all u. By the induction hypothesis $\{f(e')\}_\Theta(u,a) \simeq \{e'\}^{\mathcal{L}}(u,a)$ for all u. To find the value of $G(\lambda u\{e'\}^{\mathcal{L}}(u,a))$ we must see if $\lambda u\{e'\}^{\mathcal{L}}(u,a) = f_{xy}$ for some x,y, and if this is true find the value of $G(f_{xy})$. We do this in five questions. Notice that by the construction of f_{xy} $\lambda u\{e'\}^{\mathcal{L}}(u,a) \neq f_{xy}$ if $|y|^{\mathcal{L}} \geq \mu$.

First question: Are there x,y such that $|y|^{\mathcal{L}} < \mu$ and

$\lambda u\{e'\}^{\mathcal{L}}(u,a) = f_{xy}$?

No: $G(\lambda u\{e'\}^{\mathcal{L}}(u,a)) \simeq 0$,
Yes: Go on to the second question.

Second question: Let $\tau < \mu$ be the ordinal such that for some x and y: $\tau = |y|^{\mathcal{L}}$ and $\lambda u\{e'\}^{\mathcal{L}}(u,a) = f_{xy}$.
Is there an ordinal $\nu < \lambda$ such that $\sigma_\rho < \tau$ when $\rho < \nu$, and $\sigma_\nu \geq \tau$?

No : $G(\lambda u\{e'\}^{\mathcal{L}}(u,a)) \simeq 0$,
Yes: Go on to the third question.

Third question: Let ν, τ be as above. Is there an x such that $|x|_\Theta \leq \eta_\nu$ and $\lambda u\{e'\}^{\mathcal{L}}(u,a) = f_{xy}$, where $|y|^{\mathcal{L}} = \tau$?

No : $G(\lambda u\{e'\}^{\mathcal{L}}(u,a)) \simeq 0$,
Yes: Go on to the fourth question.

Fourth question: Is τ $\mathcal{L}$-subconstructive ?

Yes: $G(\lambda u\{e'\}^{\mathcal{L}}(u,a)) \simeq 1$,
No : Go on to the fifth question.

Fifth question: Let $\epsilon = \eta_\nu - \sup\{\eta_\rho : \rho < \nu\}$. Is there an ordinal π such that $\tau < \pi \leq \tau + \epsilon$ and π is $\mathcal{L}^\circ$-subconstructive ?

Yes: $G(\lambda u\{e'\}^{\mathcal{L}}(u,a)) \simeq 0$,
No : $G(\lambda u\{e'\}^{\mathcal{L}}(u,a)) \simeq 1$.

Next we examine the first two questions to find Θ-indexes for them. The examination of the last three questions will be omitted.

First question. $|y|^{\mathcal{L}} < \mu \Longleftrightarrow \exists u(|y|^{\mathcal{L}} \leq |\{e'\}^{\mathcal{L}}(u,a)|)$. By the induction hypothesis $|y|^{\mathcal{L}} < \mu \Longleftrightarrow \exists u\, p(y, \langle e', u, a\rangle) \simeq 0$. $\lambda u\, p(y, \langle e', u, a\rangle)$ is total. Hence the quantifier $\exists u$ can be expressed by E, and the relation "$|y|^{\mathcal{L}} < \mu$" is Θ-computable, uniformly in e, a. Let e_1 be a Θ-index for the characteristic function of this relation. To describe f_{xy} we need all information about the $\mathcal{L}$-computations of length $< |y|^{\mathcal{L}}$. By the induction hypothesis this information can be obtained from $\lambda e\, a\ \{f(e)\}_{\Theta}(a)$ and p when $|y|^{\mathcal{L}} < \mu$. Hence there is an index e_2 such that $f_{xy} = \lambda u\{e_2\}_{\Theta}(u,x,y,f(e),a)$ when $|y|^{\mathcal{L}} < \mu$. In the first question we ask if the following statement is true: $\exists x \exists y(|y|^{\mathcal{L}} < \mu$ and $f_{xy} = \lambda u\{f(e')\}_{\Theta}(u,a))$. A Θ-index for the relation inside the brackets can be found from e_1 and e_2. The quantifiers $\exists x \exists y$ can be expressed by E.

Second question: Since $\mu < \sigma$ the answer to this question is yes. So to compute $G(\lambda u\{e'\}^{\mathcal{L}}(u,a))$ go on to question 3.

Description of $p(x,y)$. Suppose $\inf(|x|^{\mathcal{L}}, |y|^{\mathcal{L}}) = \mu$. The definition of $p(x,y)$ is by cases. The form of x and y determines which case we are in. Only one case will be studied here: When both x and y correspond to substitutions, i.e. $x = \langle\langle 10,n,g,h\rangle,a\rangle$, $y = \langle\langle 10,n',g',h'\rangle,a'\rangle$. If x is convergent then the immediate subcomputations of x are $\langle g,a\rangle$ and $\langle h,\{g\}^{\mathcal{L}}(a),a\rangle$. If y is convergent then the immediate subcomputations of y are $\langle g',a'\rangle$ and $\langle h',\{g'\}^{\mathcal{L}}(a'),a'\rangle$.

$$
\begin{aligned}
p(x,y) \simeq 0 \quad &\text{if} \quad |\{g\}^{\mathcal{L}}(a)|,\ |\{h\}^{\mathcal{L}}(\{g\}^{\mathcal{L}}(a),a)| \leq |\{g'\}^{\mathcal{L}}(a')| \\
&\text{or} \quad |\{g\}^{\mathcal{L}}(a)|,\ |\{h\}^{\mathcal{L}}(\{g\}^{\mathcal{L}}(a),a)| \leq |\{h'\}^{\mathcal{L}}(\{g'\}^{\mathcal{L}}(a'),a')| \\
\simeq 1 \quad &\text{if} \quad |\{g\}^{\mathcal{L}}(a)| > |\{g'\}^{\mathcal{L}}(a')|,\ |\{h'\}^{\mathcal{L}}(\{g'\}^{\mathcal{L}}(a'),a')| \\
&\text{or} \quad |\{h\}^{\mathcal{L}}(\{g\}^{\mathcal{L}}(a),a)| > |\{g'\}^{\mathcal{L}}(a')|,\ |\{h'\}^{\mathcal{L}}(\{g'\}^{\mathcal{L}}(a'),a')|.
\end{aligned}
$$

"$|\{g\}^{\mathcal{L}}(a)|,\ |\{h\}^{\mathcal{L}}(\{g\}^{\mathcal{L}}(a),a)| \leq |\{g'\}^{\mathcal{L}}(a')|$" can be expressed as: "$p(\langle g,a\rangle,\langle g',a'\rangle) \simeq 0$ and $p(\langle h,\{f(g)\}_{\Theta}(a),a\rangle,\langle g',a\rangle) \simeq 0$". If $|x|^{\mathcal{L}} \leq |y|^{\mathcal{L}}$ then by the induction hypothesis $p(\langle g,a\rangle,\langle g',a'\rangle)$ is defined, $\{f(g)\}_{\Theta}(a) \simeq \{g\}^{\mathcal{L}}(a)$, $\{g\}^{\mathcal{L}}(a)\downarrow$, and $p(\langle h,\{f(g)\}_{\Theta}(a),a\rangle,\langle g',a'\rangle)$ is defined. The other parts of the definition of p can be replaced by similar expressions. This describes $p(x,y)$ in terms of $p(x',y')$, where $\inf(x',y') < \mu$, and $\{f(e')\}_{\Theta}(a')$ where $|\{e'\}^{\mathcal{L}}(a')| < \mu$.

This proves a). Note that $\{f(e)\}_{\Theta}(a) \simeq \{e\}^{\mathcal{L}}(a)$ may not be true if $|\{e\}^{\mathcal{L}}(a)| \geq \sigma$, because in this case the answer to question 2 can be no. In the construction of f we assumed that the answer was yes. Because of this $p(x,y)$ may not have the right value when $\inf(|x|^{\mathcal{L}}, |y|^{\mathcal{L}}) > \sigma$. For f occurs in the description of $p(x,y)$. In the case given above f occurs in the expression $\{f(g)\}_{\Theta}(a)$.

Proof of b). Θ-indexes for f' and p' can be found by the second recursion theorem. The construction is similar to the construction in a). Let $\mu < \varkappa$. Suppose $\{f'(e)\}_{\Theta}(a,P) \simeq \{e\}^{\mathcal{L}}(a)$ for all e,a such that $|\{e\}^{\mathcal{L}}(a)| < \mu$, and $p'(x,y,P)$ is defined and has the right value when $\inf(|x|^{\mathcal{L}}, |y|^{\mathcal{L}}) < \mu$. As in a) we describe $\{f'(e)\}_{\Theta}(a,P)$ when $|\{e\}^{\mathcal{L}}(a)| = \mu$, and $p'(x,y,P)$ when $\inf(|x|^{\mathcal{L}}, |y|^{\mathcal{L}}) = \mu$. $p'(x,y,P)$ is defined from f' in the same way as p was defined from f. So it is enough to regard $\{f'(e)\}_{\Theta}(a,P)$ where $|\{e\}^{\mathcal{L}}(a)| = \mu$. As in a) we only regard the case $\{e\}^{\mathcal{L}}(a) \simeq G(\lambda u \{e'\}^{\mathcal{L}}(u,a))$. The description of $\{f'(e)\}_{\Theta}(a,P)$ is the same as the description of $\{f(e)\}_{\Theta}(a)$ in a) up to the second question. Here the descriptions differ, as we cannot assume that the answer is yes. Below follows the examination of the second question.

Let $Q = \{r : r \in S$ and r is a convergent $\mathcal{L}$-computation$\}$. If $r \in Q$ let $|r|_Q$ be the ordinal ν such that $|r|^{\mathcal{L}} = \sigma_\nu$. Hence $|r|_Q$ is the order type of the $\mathcal{L}$-subconstructive ordinals $< |r|^{\mathcal{L}}$. If $r \notin Q$ let $|r|_Q$ be the order type of the $\mathcal{L}$- subconstructive ordinals. The set P was defined in the beginning of the proof of the theorem. If $r \in P$ let $|r|_P$ be the ordinal ν such that $|r|_{\Theta} = \eta_\nu$. If $r \notin P$ let $|r|_P = \lambda$. From the result in a) one can deduce that there is an index e_1 such that if $|s|_Q < \lambda$, $|s|_Q \leq |s'|_Q$ then $\{e_1\}_{\Theta}(s,s') \simeq 0$, if $|s'|_Q < \lambda$, $|s|_Q > |s'|_Q$ then $\{e_1\}_{\Theta}(s,s') \simeq 1$. From e_1 one can construct an index e_2 such that if $r \in P$, $|r|_P \leq |s|_Q$ then $\{e_2\}_\Theta(r,s) \simeq 0$, if $|r|_P > |s|_Q$ then $\{e_2\}_\Theta(r,s) \simeq 1$. ($e_1$ and e_2 can be found by applications of the second recursion theorem.)

The following is an equivalent reformulation of the second question: Is there $r \in P$ such that $|r|_P$ is the order type of the $\mathcal{L}$-subconstructive ordinals $< |y|^{\mathcal{L}}$? A Θ-index for "$r \in P$ and $|r|_P$ is the order type of the $\mathcal{L}$-subconstructive ordinals $< |y|^{\mathcal{L}}$" can be found from the following instructions: First see if $r \in P$. If not give output 1. If $r \in P$ let ν be the order type of the $\mathcal{L}$-subconstructive ordinals $< |y|^{\mathcal{L}}$. It remains to decide whether or not $|r|_P = \nu$. $|r|_P \leq \nu \Longleftrightarrow \forall r'[|r'|_P < |r|_P \Rightarrow \exists s(|s|^{\mathcal{L}} < |y|^{\mathcal{L}}$ and $|r'|_P \leq |s|_Q)]$. By the induction hypothesis $\lambda s(1 - p'(y,s,P))$ is the characteristic function of "$|s|^{\mathcal{L}} < |y|^{\mathcal{L}}$" as a relation of s. $\lambda s\{e_2\}_\Theta(r',s)$ is the characteristic function of "$|r'|_P \leq |s|_Q$" as a relation of s when $r' \in P$. The quantifiers $\forall r'$, $\exists s$ can be expressed by the functional E. This proves that there is an index e_3, such that $\lambda r\{e_3\}_\Theta(r,y,P)$ is the characteristic function of $|r|_P \leq \nu$. In a similar way one can decide whether or not $|r|_P \geq \nu$. Hence there is an index e_4 such that $\{e_4\}_\Theta(y,P)$ gives the answer to the second question. □ (prop. 4)

<u>Remark 1</u>: In the examination of question 2 we ask whether or not $r \in P$. Hence P is used negatively. This is the only place where it is necessary to regard P as an argument in the computation. In all other parts of the construction P is used positively. Hence expressions like $r \in P$ can be replaced by $\{e\}_\Theta(r)\downarrow$, where e is a Θ-index for P.

<u>Remark 2</u>: Let $\lambda' \leq \lambda$. Let $P' = P_{\lambda'} = \{r \in P : |r|_P < \lambda'\}$. Let $\eta' = \sup\{\eta_\rho + 1 : \rho < \lambda'\}$. If we replace P by P' in the construction of G we will obtain another functional G'. Let $\mathcal{L}' = G', E$. The

constructions of G and G' are equal up to stage $\sigma' = \sup\{\sigma_\rho + 1 : \rho < \lambda'\}$, i.e. $G'_\mu = G_\mu$ for $\mu < \sigma'$. The functions f', p' from part b) of proposition 4 have the properties: If $\{e\}^{\mathcal{L}'}(a)\downarrow$ then $\{f'(e)\}_\Theta(a,P') \simeq \{e\}^{\mathcal{L}'}(a)$. If $x \in C^{\mathcal{L}'}$ and $|x|^{\mathcal{L}'} \leq |y|^{\mathcal{L}'}$ then $p'(x,y,P') \simeq 0$, if $|x|^{\mathcal{L}'} > |y|^{\mathcal{L}'}$ then $p'(x,y,P') \simeq 1$.

Proposition 5: The order type of the $\mathcal{L}$-subconstructive ordinals is λ.

Proof: By proposition 2 the order type $\geq \lambda$. Suppose that the order type $> \lambda$. Then there is $s_0 \in Q$ such that $|s_0|^{\mathcal{L}} = \sigma_\lambda$. By reflection we will prove that $|s_0|^{\mathcal{L}} < \sigma_\lambda$, and hence obtain a contradiction. Below follow some technical preliminaries in order to reflect.

There are indexes e_1, e_2, e_3, e_4 such that $\{e_1\}_\Theta(s_0, P')\downarrow$ iff s_0 is a convergent $\mathcal{L}'$-computation and λ' is the order type of the $\mathcal{L}'$-subconstructive ordinals $< |s_0|^{\mathcal{L}'}$. $\{e_2\}_\Theta(s_0,P')\downarrow$ iff s_0 is a convergent $\mathcal{L}'$-computation, in which case $|s_0|^{\mathcal{L}'} < |\{e_2\}_\Theta(s_0,P')|_\Theta$. $\{e_3\}_\Theta(P')\downarrow$ and $\eta' \leq |\{e_3\}_\Theta(P')|_\Theta$. $\{e_4\}_\Theta(s_0,P')\downarrow$ iff s_0 is a convergent $\mathcal{L}'$-computation, in which case $\eta' + |s_0|^{\mathcal{L}'} < |\{e_4\}_\Theta(s_0,P')|_\Theta$. ($P', \lambda', \eta', \mathcal{L}'$ are as in remark 2 above.)

e_1 can be constructed in the following way: The statement "s_0 is a convergent $\mathcal{L}'$-computation" is Θ-semicomputable in P'. An index for this statement can be found from an index for p' in part b) of proposition 4. The statement "λ' is the order type of the $\mathcal{L}'$-subconstructive ordinals $< |s_0|^{\mathcal{L}'}$" can be reformulated by: $\forall r \in P' \exists s(|r|_P = |s|_Q$ and $|s|^{\mathcal{L}'} < |s_0|^{\mathcal{L}'})$ and $\forall s(|s|^{\mathcal{L}'} < |s_0|^{\mathcal{L}'} \Rightarrow \exists r \in P'(|r|_P = |s|_Q))$. By an application of the second recursion theorem and of the result in part a) of proposition 4 it can be proved that the relation "$r \in P$ and $s \in Q$ and $|r|_P = |s|_Q$" is Θ-semicomputable. This relation can be used to express "$|r|_P = |s|_Q$" in the statement above. The other parts of the statement can be expressed by the function p' in part b) of proposition 4. The quantifiers can be expressed by E.

e_2 can be constructed in the following way: Let $s_0 = \langle e,a \rangle$. Then $s_0 \in C^{\mathcal{L}'}$ iff $\{e\}^{\mathcal{L}'}(a)\downarrow$ iff $\{f'(e)\}_\Theta(a,P')\downarrow$. The function f' in proposition 4 b) can be chosen such that the following is true: If $\{e\}^{\mathcal{L}'}(a)\downarrow$ then $|\{e\}^{\mathcal{L}'}(a)|^{\mathcal{L}'} < |\{f'(e)\}_\Theta(a,P')|_\Theta$. Let $\{e_2\}_\Theta(s_0,P') \simeq \{f'(e)\}_\Theta(a,P')$.

Let e_3 be an index for the following instruction: For all $r \in P'$ compute the computation r. Then $\{e_3\}_\Theta(P')\downarrow$, and $|r|_\Theta < |\{e_3\}_\Theta(P')|_\Theta$ for all $r \in P'$. Hence $\eta' \leq |\{e_3\}_\Theta(P')|_\Theta$.

From e_2 and e_3 one can construct an index e_4 such that $\{e_4\}_\Theta(s_0,P')\downarrow$ iff $\{e_2\}_\Theta(s_0,P')\downarrow$ and $\{e_3\}_\Theta(P')\downarrow$, in which case $|\{e_3\}_\Theta(P')|_\Theta + |\{e_2\}_\Theta s_0,P')|_\Theta < |\{e_4\}_\Theta(s_0,P')|_\Theta$. Hence $\{e_4\}_\Theta(s_0,P')\downarrow$ iff s_0 is a convergent $\mathcal{L}'$-computation, in which case $\eta' + |s_0|^{\mathcal{L}'} < |\{e_4\}_\Theta(s_0,P')|_\Theta$.

Now we come to the reflecting statement. It is a conjunction of three parts a), b) and c).

a) $\exists\lambda' \leq \lambda \ (x = P_{\lambda'})$.

b) s_0 is a convergent $\mathcal{L}'$-computation, and λ' is the order type of the $\mathcal{L}'$-subconstructive ordinals $< |s_0|^{\mathcal{L}'}$.

c) If $r \notin x$ then $\eta' + |s_0|^{\mathcal{L}'} < |r|_\Theta$.

There is an index e_5 such that if $\{e_5\}_\Theta(s_0,x)\downarrow$ then a), b) and c) are satisfied for x. e_5 can be constructed as follows: a) can be expressed by "$x \subseteq S$, and each element in x is a convergent Θ-computation, and for all r,r': If $r \in x$ and $|r'|_\Theta \leq |r|_\Theta$ then $r' \in x$". A Θ-index for this statement can be found. If $x = P_{\lambda'} = P'$ then b) is satisfied iff $\{e_1\}_\Theta(s_0,x)\downarrow$. The following statement implies c) when $x = P'$: "for all $r \notin x$ $|\{e_4\}_\Theta(s_0,x)| < |r|_\Theta$". A Θ-index for this statement can be found. Let e_5 be a Θ-index for the conjunction of these statements.

Let B be defined by: $x \in B \Leftrightarrow \{e_5\}_\Theta(s_0,x)\downarrow$. Then B is Θ-semicomputable in s_0, and by a) in the reflecting statement each element in B is a subset of S (in fact $x \in B \Rightarrow x = P_\nu$ for a $\nu \leq \lambda$). $P \in B$, for a) is trivially satisfied for $x = P$. By assumption b) is satisfied for $x = P$. To see that c) is satisfied for $x = P$ let $r \notin P$. Then r is a divergent Θ-computation, and $|r|_\Theta = \kappa_\Theta$. Now $\{e_4\}_\Theta(s_0,x)\downarrow$, hence $|\{e_4\}_\Theta(s_0,x)|_\Theta < |r|_\Theta$, and $\{e_5\}_\Theta(s_0,x)\downarrow$. P is Θ-semicomputable. By the corollary of theorem 8 there is an element x_o in B which is Θ-computable in s_0. Since P is not Θ-computable in s_0 it follows that $x_o = P_{\lambda'}$ where $\lambda' < \lambda$. Let $P' = P_{\lambda'}$.

Since the reflecting statement is true for P': s_0 is a convergent $\mathcal{L}'$-computation, and λ' is the order type of the $\mathcal{L}'$-subconstructive ordinals $< |s_0|^{\mathcal{L}'}$, and if $r \notin P'$ then $\eta' + |s_0|^{\mathcal{L}'} < |r|_\Theta$.

To prove that $|s_0|^{\mathcal{L}} < \lambda$ we go back to the constructions of G and G'. As mentioned in remark 2 $G'_\mu = G_\mu$ when $\mu < \sigma'$ where $\sigma' = \sup\{\sigma_\rho + 1 : \rho < \lambda'\}$. Let $\tau = \sigma'$. The $\mathcal{L}$-subconstructive and the $\mathcal{L}'$-subconstructive ordinals $< \tau$ are the same, and the order type of these ordinals is λ'. τ is $\mathcal{L}$-subconstructive iff τ is $\mathcal{L}'$-subconstructive. By b) in the reflecting statement $|s_0|^{\mathcal{L}'}$ is the λ'-th $\mathcal{L}'$-subconstructive ordinal.

<u>Claim</u>: $G'_\mu = G_\mu$ for all $\mu < |s_0|^{\mathcal{L}'}$.

Proof: If $\tau = |s_0|^{\mathcal{L}'}$ (which is the case of τ is $\mathcal{L}$-subconstructive) then the claim is true by the discussion above. Suppose $\tau < |s_0|^{\mathcal{L}'}$. This is the case if τ is not $\mathcal{L}$-subconstructive. In the construction of G'_τ we are in case 2 for the first time. Hence

$$G'(f) = \begin{cases} (\bigcup_{\mu<\tau} G_\mu)(f) & \text{if } \exists\mu < \tau \quad f \in \operatorname{dom} G_\mu \\ 0 & \text{otherwise.} \end{cases}$$

In the construction of G_τ we are in case 1 since the order type of the $\mathcal{L}$-subconstructive ordinals less than τ is λ', and $\lambda' < \lambda$. We are in subcase II, because τ is not $\mathcal{L}$-subconstructive by the assumption $\tau < |s_0|^{\mathcal{L}'}$. The functional G^o mentioned at this stage of the construction is the same as G'. Hence s_0 is a convergent $\mathcal{L}^o$-computation, $|s_0|^{\mathcal{L}'} = |s_0|^{\mathcal{L}^o}$, and $|s_0|^{\mathcal{L}^o}$ is the first ordinal $> \tau$ which is $\mathcal{L}^o$-subconstructive. Let $\epsilon = \eta_{\lambda'} - \sup\{\eta_\rho : \rho < \lambda'\}$. By c) in the reflecting statement $\eta' + |s_0|^{\mathcal{L}^o} < |r|_\Theta$ if $r \notin P'$, i.e. if $|r|_\Theta \geq \eta_{\lambda'}$. Hence $\eta' + |s_0|^{\mathcal{L}^o} < \eta_{\lambda'}$. By definition $\eta' = \sup\{\eta_\rho + 1 : \rho < \lambda'\}$. Hence $\sup\{\eta_\rho : \rho < \lambda'\} + |s_0|^{\mathcal{L}^o} < \eta_{\lambda'}$.

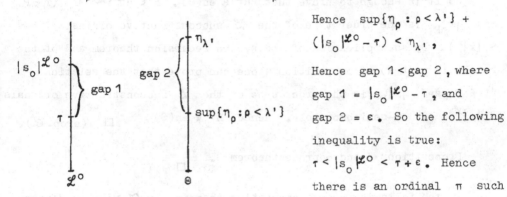

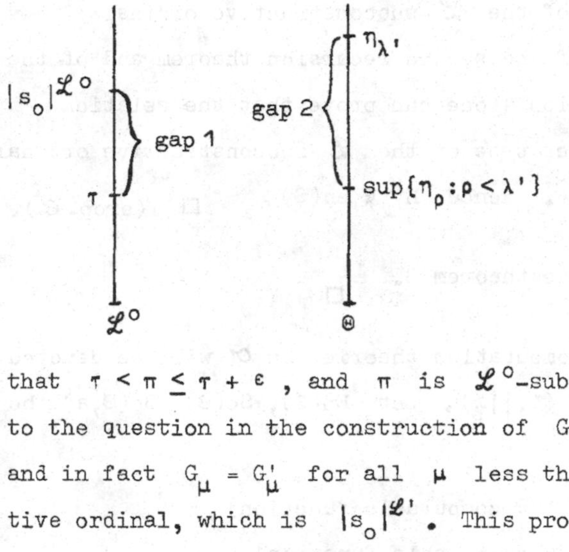

Hence $\sup\{\eta_\rho : \rho < \lambda'\} + (|s_0|^{\mathcal{L}^o} - \tau) < \eta_{\lambda'}$,

Hence gap 1 < gap 2, where gap 1 $= |s_0|^{\mathcal{L}^o} - \tau$, and gap 2 $= \epsilon$. So the following inequality is true: $\tau < |s_0|^{\mathcal{L}^o} < \tau + \epsilon$. Hence there is an ordinal π such that $\tau < \pi \leq \tau + \epsilon$, and π is $\mathcal{L}^o$-subconstructive. So the answer to the question in the construction of G_τ is yes. Hence $G_\tau = G'_\tau$, and in fact $G_\mu = G'_\mu$ for all μ less than the next $\mathcal{L}^o$-subconstructive ordinal, which is $|s_0|^{\mathcal{L}'}$. This proves the claim.

By the claim the $\mathcal{L}'$-computations of length less than $|s_0|^{\mathcal{L}'}$ are identical to the $\mathcal{L}$-computations of length less than $|s_0|^{\mathcal{L}'}$. If $s_0 = \langle e,a \rangle$ then $\{e\}^{\mathcal{L}'}(a)\downarrow$. All immediate subcomputations $\{e'\}^{\mathcal{L}'}(a')$ of $\{e\}^{\mathcal{L}'}(a)$ have length less than $|s_0|^{\mathcal{L}'}$, hence $\{e'\}^{\mathcal{L}'}(a') \simeq \{e'\}^{\mathcal{L}}(a')$, and $|\{e'\}^{\mathcal{L}}(a')|^{\mathcal{L}} < |s_0|^{\mathcal{L}'}$. Hence $\{e\}^{\mathcal{L}}(a)\downarrow$, and $|\{e\}^{\mathcal{L}}(a)|^{\mathcal{L}} = |s_0|^{\mathcal{L}'}$, i.e. $|s_0|^{\mathcal{L}} = |s_0|^{\mathcal{L}'} = \sigma_{\lambda'}$, a contradiction. This proves proposition 5.

□ (prop. 5)

Proposition 6: $S-en(\mathcal{L}) \subseteq S-en(\Theta)$; and for all $r \in S$: $sc(\mathcal{L},r) \subseteq sc(\Theta,r)$.

Proof: Let $H^{\mathcal{L}} = \{\langle e,a\rangle : \{e\}^{\mathcal{L}}(a)\downarrow$ and $|\{e\}^{\mathcal{L}}(a)|^{\mathcal{L}} < \sigma\}$. By proposition 5 σ is the supremum of the $\mathcal{L}$-subconstructive ordinals. If $X \subseteq S$, $X \in S-en(\mathcal{L})$ then there is an index e such that for all r: $r \in X \iff \langle e,r\rangle \in H^{\mathcal{L}}$.

Suppose $X \subseteq I$, $X \in sc(\mathcal{L},r)$, $r \in S$. By the method used in the proof of proposition 3 one can prove that there are indexes e_1, e_2 such that for all x: $x \in X \iff \langle e_1,x,r\rangle \in H^{\mathcal{L}}$, $x \notin X \iff \langle e_2,x,r\rangle \in H^{\mathcal{L}}$.

So it is enough to prove that $H^{\mathcal{L}} \in en(\Theta)$. $x \in H^{\mathcal{L}} \iff \exists r$ ($r \in P$ and $|r|_P$ is the order type of the $\mathcal{L}$-subconstructive ordinals $< |x|^{\mathcal{L}}$). By an application of the second recursion theorem and of the result in part a) of proposition 4 one can prove that the relation "$r \in P$ and $|r|_P$ is the order type of the $\mathcal{L}$-subconstructive ordinals $< |x|^{\mathcal{L}}$" is Θ-semicomputable. Hence $H^{\mathcal{L}} \in en(\Theta)$. □ (prop. 6)

Propositions 3 and 6 prove theorem 9. □

In the following pages computation theories on $\mathfrak{A}$ will be denoted by Θ, Ψ instead of $(\Theta,||_{\Theta}), (\Psi,||_{\Psi})$. Let $En(\Theta)$, $Sc(\Theta)$, $Sc(\Theta,a)$ be defined by:

$En(\Theta) = \{\varphi : \varphi$ is a partial Θ-computable function$\}$,

$Sc(\Theta) = \{f : f$ is a total Θ-computable function$\}$,

$Sc(\Theta,a) = \{f : f$ is total and Θ-computable in $a\}$.

Let Ψ, Θ be normal computation theories on $\mathfrak{A}$. Let $\sim, \leq, <_1, <_2$ be defined by:

$\Psi \sim \Theta \iff \mathrm{En}(\Psi) = \mathrm{En}(\Theta)$,

$\Psi \leq \Theta \iff \mathrm{En}(\Psi) \subseteq \mathrm{En}(\Theta)$,

$\Psi <_1 \Theta \iff \mathrm{En}(\Psi) \subseteq \mathrm{En}(\Theta)$ and $\exists x(\kappa_\Psi^x < \kappa_\Theta^x)$,

$\Psi <_2 \Theta \iff \mathrm{En}(\Psi) \subsetneqq \mathrm{En}(\Theta)$.

<u>Lemma 26</u>: Let Ψ and Θ be normal.

a) $\Psi \leq \Theta \Rightarrow \forall a(\mathrm{Sc}(\Psi,a) \subseteq \mathrm{Sc}(\Theta,a))$ and $\forall a(\kappa_\Psi^a \leq \kappa_\Theta^a)$.

b) $\Psi \leq \Psi'$, $\Psi' <_1 \Theta'$, $\Theta' \leq \Theta \Rightarrow \Psi <_1 \Theta$,

c) $\Psi <_1 \Theta \Rightarrow \Psi <_2 \Theta$.

Proof: a) Suppose $\Psi \leq \Theta$. Let $f \in \mathrm{Sc}(\Psi,a)$. There is an index e such that $f = \lambda x\{e\}_\Psi(x,a)$. Let $\varphi = \lambda xa\{e\}_\Psi(x,a)$. $\varphi \in \mathrm{En}(\Psi)$. By assumption $\varphi \in \mathrm{En}(\Theta)$. Hence there is an index e' such that $\varphi = \lambda x a\{e'\}_\Theta(x,a)$. Hence $f = \lambda x\{e'\}_\Theta(x,a)$, i.e. $f \in \mathrm{Sc}(\Theta,a)$.

$\kappa_\Psi^a = \sup\{\tau : \tau$ is the length of a prewellordering with domain I which is Ψ-computable in $a\}$. If X is a prewellordering which is Ψ-computable in a then by the above X is Θ-computable in a. Hence $\kappa_\Psi^a \leq \kappa_\Theta^a$.

b) Suppose $\Psi \leq \Psi'$, $\Psi' <_1 \Theta'$, $\Theta' \leq \Theta$. Then $\mathrm{En}(\Psi) \subseteq \mathrm{En}(\Theta)$. Since $\Psi' <_1 \Theta'$ there is an x such that $\kappa_{\Psi'}^x < \kappa_{\Theta'}^x$. By a) $\kappa_\Psi^x \leq \kappa_{\Psi'}^x$ and $\kappa_{\Theta'}^x \leq \kappa_\Theta^x$. Hence $\kappa_\Psi^x < \kappa_\Theta^x$. Hence $\Psi <_1 \Theta$.

c) Suppose $\Psi <_1 \Theta$. Then $\mathrm{En}(\Psi) \subseteq \mathrm{En}(\Theta)$. If $\mathrm{En}(\Psi) = \mathrm{En}(\Theta)$ then by a) $\kappa_\Psi^x = \kappa_\Theta^x$ for all x, a contradiction. Hence $\mathrm{En}(\Psi) \subsetneqq \mathrm{En}(\Theta)$. □

Let $\mathcal{L}$ be a list, and let Θ be a computation theory. $\mathcal{L}$ is <u>Θ-computable</u> if each function and relation in $\mathcal{L}$ is Θ-computable, and each functional in $\mathcal{L}$ is weakly Θ-computable. Let $\mathcal{P}(\mathcal{L})$ denote the computation theory obtained from $\mathcal{L}$.

<u>Lemma 27</u>: a) If $\mathcal{L}$ is Θ-computable then $\mathcal{P}(\mathcal{L}) \leq \Theta$.
b) If Θ is normal and $\mathcal{P}(\mathcal{L}) \sim \Theta$ then $\mathcal{L}$ is Θ-computable.

Proof: a) Suppose $\mathcal{L}$ is Θ-computable. One can prove that there is a Θ-computable total function f such that for all e, a: $\{e\}^{\mathcal{L}}(a) \simeq \{f(e)\}_{\Theta}(a)$. Hence $\mathcal{P}(\mathcal{L}) \leq \Theta$.

b) Suppose Θ is normal and $\mathcal{P}(\mathcal{L}) \sim \Theta$. Then the functions and the relations in the list $\mathcal{L}$ are Θ-computable. Let F be a functional in the list $\mathcal{L}$. We must prove that F is weakly Θ-computable. We give instructions how to compute $F(\lambda x\{e\}_{\Theta}(x,a))$. Let $n \geq 0$, and let a range over all lists of length n. Let $\varphi_{\Theta}(e,x,a) \simeq \{e\}_{\Theta}(x,a)$. Then $\varphi_{\Theta} \in \mathrm{En}(\Theta)$, hence $\varphi_{\Theta} \in \mathrm{En}(\mathcal{L})$. Let $f(n)$ be an $\mathcal{L}$-index for φ_{Θ}. Then $\lambda x\{e\}_{\Theta}(x,a) = \lambda x\{f(n)\}^{\mathcal{L}}(e,x,a)$. $F(\lambda x\{e\}_{\Theta}(x,a)) \simeq F(\lambda x\{f(n)\}^{\mathcal{L}}(e,x,a)) \simeq \{g(n)\}^{\mathcal{L}}(e,a)$, where $g(n)$ can be constructed from $f(n)$ in a primitive recursive way. Let $\varphi_{\mathcal{L}}(e,a) \simeq \{e\}^{\mathcal{L}}(a)$. Then $\varphi_{\mathcal{L}} \in \mathrm{En}(\mathcal{L})$, hence $\varphi_{\mathcal{L}} \in \mathrm{En}(\Theta)$. So $F(\lambda x\{e\}_{\Theta}(x,a)) \simeq \varphi_{\mathcal{L}}(g(n),e,a)$. There is a Θ-computable total function h such that $F(\lambda x\{e\}_{\Theta}(x,a)) \simeq \{h(n)\}_{\Theta}(e,a)$. To conclude that F is weakly Θ-computable the following should be true: $|\{h(n)\}_{\Theta}(e,a)|_{\Theta} > |\{e\}_{\Theta}(x,a)|_{\Theta}$ for all x. This can be obtained as follows: Let $E(\lambda x\{e\}_{\Theta}(x,a))$ be a subcomputation of $\{h(n)\}_{\Theta}(e,a)$. Since E is weakly Θ-computable we get the right ordinal inequalities. □

<u>Lemma 28</u>: Suppose Θ is normal, $\mathcal{L}$ is Θ-computable and $\forall x(\kappa^{x}_{\mathcal{L}} = \kappa^{x}_{\Theta})$. Then there is a normal list $\mathcal{L}'$ such that $\mathcal{P}(\mathcal{L}') \sim \Theta$.

Proof: There is a Θ-index e_1 such that $\{e_1\}_{\Theta}(e,y)\downarrow$ iff $\lambda x\{e\}_{\Theta}(x,y)$ is total, in which case $|\{e\}_{\Theta}(x,y)| < |\{e_1\}_{\Theta}(e,y)|$ for all x. (e_1 can for instance be an index for the computation $E(\lambda x\{e\}_{\Theta}(x,y))$.)

Let f be a variable for total unary functions $I \to S$. Let $\mathrm{Ord}(f)$ be the least ordinal τ such that for some $e,y : f =$

$\lambda x\{e\}_\Theta(x,y)$, and $|\{e_1\}_\Theta(e,y)| = \tau$, if such an ordinal exists. Ord(f) is undefined otherwise.

Let G be defined by: If Ord(f) is defined then

$$G(\langle f,e,a\rangle) = \begin{cases} \{e\}_\Theta(a)+1 & \text{if } |\{e\}_\Theta(a)|_\Theta \leq \mathrm{Ord}(f) \\ 0 & \text{otherwise} \end{cases}$$

where $\langle f,e,a\rangle = \lambda x\langle f(x),e,a\rangle$. (We make the following convention: If $r \notin N$ then $r+1 = r$.) $G(g) = 0$ if g is not of the form $\langle f,e,a\rangle$, or if $g = \langle f,e,a\rangle$ and Ord(f) is undefined.

G is weakly Θ-computable. For let $\varphi = \lambda x\{e'\}_\Theta(x,a')$. To compute $G(\varphi)$ first check if φ is total. (This can for instance be done by computing $E(\varphi)$.) If φ is total check if $\varphi = \langle f,e,a\rangle$ for some f,e,a. If not let $G(\varphi) = 0$. I $\varphi = \langle f,e,a\rangle$ then Ord(f) is defined since φ is Θ-computable. It remains to see whether or not $|\{e\}_\Theta(a)|_\Theta \leq \mathrm{Ord}(f)$. Let $f = \lambda x\{e_2\}_\Theta(x,\langle a'\rangle)$.
$|\{e\}_\Theta(a)|_\Theta \leq \mathrm{Ord}(f) \iff \forall h \in N \forall y$
$([\forall x(\{h\}_\Theta(x,y) \simeq \{e_2\}_\Theta(x,\langle a'\rangle))$ and $|\{e_1\}_\Theta(h,y)| \leq |\{e_1\}_\Theta(e_2,\langle a'\rangle)|]$
$\Rightarrow |\{e\}_\Theta(a)| \leq |\{e_1\}_\Theta(h,y)|)$. This expression says:
$\forall h,y(f = \lambda x\{h\}_\Theta(x,y) \Rightarrow |\{e\}_\Theta(a)| \leq |\{e_1\}_\Theta(h,y)|)$. As a relation of e,a this is Θ-computable in a'. Hence G is weakly Θ-computable. Let $\mathcal{L}' = \mathcal{L},G,E,=_S$, where $=_S$ is the equality relation on S. Then $\mathcal{L}'$ is normal, and $\mathrm{En}(\mathcal{L}') \subseteq \mathrm{En}(\Theta)$ because $\mathcal{L}'$ is Θ-computable. To prove the opposite inclusion we need a proposition.

For all $y \in I$ let $\mu_y = \sup\{\mathrm{Ord}(f) : f$ is recursive in $\mathcal{L}',y\}$.

<u>Proposition 1</u>: $\mu_y = \kappa^y_\Theta$ for all y.

Proof: If f is recursive in $\mathcal{L}',y$ then f is Θ-computable in y, because $\mathcal{L}'$ is Θ-computable. Hence Ord(f) is defined, and $\mathrm{Ord}(f) < \kappa^y_\Theta$. Hence $\mu_y \leq \kappa^y_\Theta$.

$\kappa^y_{\mathcal{L}} \leq \kappa^y_{\mathcal{L}'} \leq \kappa^y_\Theta$ since $\mathcal{L}'$ extends $\mathcal{L}$ and $\mathcal{L}'$ is Θ-computable. By assumption $\kappa^y_{\mathcal{L}} = \kappa^y_\Theta$. Hence $\kappa^y_{\mathcal{L}} = \kappa^y_{\mathcal{L}'} = \kappa^y_\Theta$. By lemma 20 $\kappa^y_{\mathcal{L}'}$ is the supremum of the lengths of the prewellorderings with domain $\subseteq I$ which are recursive in $\mathcal{L}'$,y. To prove that $\mu_y = \kappa^y_\Theta$ it is enough to prove that for each prewellordering which is recursive in $\mathcal{L}'$,y there is a function f which is recursive in $\mathcal{L}'$,y such that $\mathrm{Ord}(f) \geq$ the length of the prewellordering.

So let $X \subseteq I^2$ be a prewellordering which is recursive in $\mathcal{L}'$,y. If $x \in \mathrm{dom}(X)$ let $|x|$ be the length of that part of the prewellordering which is below x. By the second recursion theorem there is an index $\hat{e}$ such that for all $x \in \mathrm{dom}(X)$ $\lambda t\{\hat{e}\}^{\mathcal{L}'}(t,x,y)$ is total, and $|x| \leq \mathrm{Ord}(\lambda t\{\hat{e}\}^{\mathcal{L}'}(t,x,y))$. The construction goes as follows. Suppose $x \in \mathrm{dom}(X)$, and $\lambda t\{\hat{e}\}^{\mathcal{L}'}(t,x',y)$ is total and $|x'| \leq \mathrm{Ord}(\lambda t\{\hat{e}\}^{\mathcal{L}'}(t,x',y))$ for all x' such that $|x'| < |x|$. Let $f_{x'} = \lambda t\{\hat{e}\}^{\mathcal{L}}(t,x',y)$ for $|x'| < |x|$. The set $B = \{\langle e,a,z\rangle : \{e\}_\Theta(a) \simeq z$, and $|\{e\}_\Theta(a)|_\Theta \leq \mathrm{Ord}(f_{x'})$ for some $|x'| < |x|\}$ is recursive in $\mathcal{L}'$,x,y, uniformly in x, by the construction of G, and since X is recursive in $\mathcal{L}'$,y. B contains the graphs of all functions f with $\mathrm{Ord}(f) \leq \mathrm{Ord}(f_{x'})$ for some $|x'| < |x|$. By asking questions about B one can see whether or not $\lambda t\{e\}_\Theta(t,u)$ is total and $\mathrm{Ord}(\lambda t\{e\}_\Theta(t,u)) \leq \mathrm{Ord}(f_{x'})$ for some $|x'| < |x|$, for any e,u. Let P(e,u) be the statement "$\lambda t\{e\}_\Theta(t,u)$ is total, and $\mathrm{Ord}(\lambda t\{e\}_\Theta(t,u)) \leq \mathrm{Ord}(f_{x'})$ for some $|x'| < |x|$". Then P is recursive in $\mathcal{L}'$,x,y, uniformly in x. By the construction of G the function φ defined by: $\varphi(\langle e,a\rangle) \simeq \{e\}_\Theta(a)$ if $|\{e\}_\Theta(a)|_\Theta \leq \mathrm{Ord}(f_{x'})$ for some $|x'| < |x|$, is partial recursive in $\mathcal{L}'$,x,y, uniformly in x. Let f_x be defined by: $f_x(\langle e,u\rangle) = 0$ if not P(e,u), $f_x(\langle e,u\rangle) \simeq \{e\}_\Theta(\langle e,u\rangle,u)+1$ if P(e,u). Then f_x is different from all functions f such that $\mathrm{Ord}(f) \leq \mathrm{Ord}(f_{x'})$ for some $|x'| < |x|$. f_x is recursive in $\mathcal{L}'$,x,y, uniformly in x. So $\mathrm{Ord}(f_{x'}) < \mathrm{Ord}(f_x)$ when $|x'| < |x|$. Hence $|x| \leq \mathrm{Ord}(f_x)$ by the induction hypothesis. Let $\Psi(\hat{e},t,x,y) \simeq f_x(t)$. Choose $\hat{e}$ such that $\Psi(\hat{e},t,x,y) \simeq \{\hat{e}\}^{\mathcal{L}'}(t,x,y)$ for all t,x,y.

In the same way one can construct a function f such that f is recursive in $\mathcal{L}', y$, and $|x| < \mathrm{Ord}(f)$ for all $x \in \mathrm{dom}(X)$. This proves the proposition. $\square$ (prop. 1)

Proposition 2: $\mathrm{En}(\Theta) \subseteq \mathrm{En}(\mathcal{L}')$.

Proof: Suppose $\{e\}_\Theta(a)\downarrow$. By proposition 1 there is an index m such that $f = \lambda t\{m\}^{\mathcal{L}'}(t,\langle e,a\rangle)$ is total, and $|\{e\}_\Theta(a)|_\Theta \leq \mathrm{Ord}(f)$. Hence $G(\langle \lambda t\{m\}^{\mathcal{L}}(t,\langle e,a\rangle),e,a\rangle) \simeq \{e\}_\Theta(a)+1$. Since $\mathcal{L}'$ is normal there is a selection function $\varphi(e,a)$ which is partial recursive in $\mathcal{L}'$ and which picks out such an m. Hence $\{e\}_\Theta(a) \simeq G(\langle \lambda t\{\varphi(e,a)\}^{\mathcal{L}}(t,\langle e,a\rangle),e,a\rangle)-1$. Hence $\mathrm{En}(\Theta) \subseteq \mathrm{En}(\mathcal{L}')$.

This proves lemma 28. $\square$

Definitions: Θ has property 1 if for all normal lists $\mathcal{L}$: if $\mathcal{L}$ is Θ-computable then $\mathcal{P}(\mathcal{L}) <_1 \Theta$. Θ has property 2 if for all normal lists $\mathcal{L}$: If $\mathcal{L}$ is Θ-computable then $\mathcal{P}(\mathcal{L}) <_2 \Theta$.

Lemma 29: Suppose Θ is normal. Then Θ has property 1 iff Θ has property 2.

Proof: By c) in lemma 26 $\Psi <_1 \Theta \Rightarrow \Psi <_2 \Theta$. Hence if Θ has property 1 then Θ has property 2.

Suppose Θ has not property 1. Then there is a normal list which is Θ-computable such that not $\mathcal{P}(\mathcal{L}) <_1 \Theta$. By a) in lemma 27 $\mathrm{En}(\mathcal{L}) \subseteq \mathrm{En}(\Theta)$. Hence $\forall x(\kappa_{\mathcal{L}}^x = \kappa_\Theta^x)$. By lemma 28 there is a normal list $\mathcal{L}'$ such that $\mathcal{P}(\mathcal{L}') \sim \Theta$. By b) of lemma 27 $\mathcal{L}'$ is Θ-computable. $\mathcal{P}(\mathcal{L}') <_2 \Theta$ is not true. Hence Θ has not property 2.

Definition: Θ is Mahlo if Θ has property 1.

Theorem 10: Let Θ be normal. Then Θ is not Mahlo iff there is a normal list $\mathcal{L}$ such that $\mathcal{P}(\mathcal{L}) \sim \Theta$.

Proof: Suppose Θ is not Mahlo. By lemma 29 Θ has not property 2. Hence there is a normal list $\mathscr{L}$ such that $\mathscr{L}$ is Θ-computable and not $\mathscr{P}(\mathscr{L}) <_2 \Theta$. Hence $\mathrm{En}(\mathscr{L}) = \mathrm{En}(\Theta)$, and $\mathscr{P}(\mathscr{L}) \sim \Theta$.

Suppose that there is a normal list $\mathscr{L}$ such that $\mathscr{P}(\mathscr{L}) \sim \Theta$. $\mathscr{L}$ is Θ-computable by b) of lemma 27. Hence Θ has not property 2. By lemma 29 Θ is not Mahlo.

□

§ 9 MORE ABOUT MAHLONESS

In ordinal recursion the notion of Mahloness is defined in the following way: An ordinal τ is Mahlo if τ is recursively regular, and all normal functions π which are τ-recursive in constants less than τ have a recursively regular fixpoint less than τ. (Definition 4.2 (b) in [1]). In § 8 the notion of Mahloness was defined in another way. The purpose of this chapter is to prove that the definition in § 8 is a natural generalization of the definition above.

To see this let us regard normal computation theories $(\Theta, ||_{\Theta})$ with domain ω, i.e. ω is strongly Θ-finite and $||_{\Theta}$ is a Θ-norm. Since the domain is ω $\kappa_{\Theta}^{x} = \kappa_{\Theta}$ for all x in the domain. We define property 1 as in § 8 : Θ has property 1 if for all normal lists $\mathscr{L}$: If $\mathscr{L}$ is Θ-computable then $\kappa^{\mathscr{L}} < \kappa_{\Theta}$.

Below we introduce ordinal recursion for $(\Theta, ||_{\Theta})$. Let $\tau \leq \kappa_{\Theta}$. Let $\Theta_{\tau} = \{(e,a,r) : \{e\}_{\Theta}(a) \simeq r$ and $|\{e\}_{\Theta}(a)|_{\Theta} < \tau\}$. Let $||_{\Theta\tau}$ be the restriction of $||_{\Theta}$ to Θ_{τ}. Then $||_{\Theta\tau}$ is a mapping of Θ_{τ} onto τ. τ is Θ-regular if $(\Theta_{\tau}, ||_{\Theta\tau})$ is a normal computation theory on ω.

Let $\tau \leq \kappa_{\Theta}$. A relation R is Θ_{τ}-semicomputable if there is an index e such that for all a : $R(a) \Longleftrightarrow |\{e\}_{\Theta}(a)| < \tau$. Let π be a partial function from τ^{n} to τ. π is Θ_{τ}-computable if the set $\{(x_1 \ldots x_n, y) : |x_1|_{\Theta}, \ldots, |x_n|_{\Theta}, |y|_{\Theta} < \tau$ and $\pi(|x_1|_{\Theta}, \ldots, |x_n|_{\Theta}) \simeq |y|_{\Theta}\}$ is Θ_{τ}-semicomputable. π is Θ-computable if π is Θ_{κ}-computable, where $\kappa = \kappa_{\Theta}$.

A function π from τ to τ is normal if it is strictly increasing and continuous at limits. An ordinal $\nu < \tau$ is a fixed point for π if $\pi(\nu) = \nu$. $\tau \leq \kappa_{\Theta}$ is Θ-Mahlo if τ is Θ-regular and all normal Θ_{τ}-computable functions have a Θ-regular fixed point less than τ. Θ is Mahlo if κ_{Θ} is Θ-Mahlo.

When $(\Theta, ||_{\Theta})$ is the computation theory obtained from a normal list $\mathcal{L}$, we will write $\mathcal{L}$-regular, $\mathcal{L}_T$-computable, $\mathcal{L}$-computable instead of Θ-regular, Θ_T-computable, Θ-computable.

In ordinary recursion theory an ordinal τ is said to be recursively regular if the following is true: If $\nu < \tau$, and π is a partial function which is τ-recursive in constants less than τ, and $\pi(\xi)\downarrow$ for all $\xi < \nu$, then there is an ordinal $\tau' < \tau$ such that $\pi(\xi) < \tau'$ for all $\xi < \nu$. (Definition 3.4 in [1]).

In view of this definition we prove the following lemma.

Lemma 30: If $(\Theta, ||_\Theta)$ is a normal computation theory on ω then the following is true: If $\nu < \varkappa_\Theta$, and π is a partial function which is Θ-computable, and $\pi(\xi)\downarrow$ for all $\xi < \nu$, then there is an ordinal $\tau' < \varkappa_\Theta$ such that $\pi(\xi) < \tau'$ for all $\xi < \nu$.

Proof: Let $\nu < \varkappa_\Theta$, and let π be a partial function which is Θ-computable, such that $\pi(\xi)\downarrow$ for all $\xi < \nu$. The set $\{(x,y): |x|_\Theta, |y|_\Theta < \varkappa_\Theta$ and $\pi(|x|_\Theta) \simeq |y|_\Theta\}$ is Θ-semicomputable. There is a Θ-computable selection function $\varphi(x)$ such that if $\exists y(\pi(|x|_\Theta) \simeq |y|_\Theta)$ then $\varphi(x)\downarrow$ and $\pi(|x|_\Theta) \simeq |\varphi(x)|_\Theta$. Choose a w such that $|w|_\Theta = \nu$. Choose an index e as follows: To compute $\{e\}_\Theta(w)$ take each x such that $|x|_\Theta < |w|_\Theta$, find $\varphi(x)$, and compute this computation, i.e. if $\varphi(x) = \langle e', a' \rangle$, compute $\{e'\}_\Theta(a')$. Then $\{e\}_\Theta(w)\downarrow$, and $|\varphi(x)|_\Theta < |\{e\}_\Theta(w)|$ for all x such that $|x|_\Theta < |w|_\Theta$. Let $\tau' = |\{e\}_\Theta(w)|$. Then $\tau' < \varkappa_\Theta$, and $\pi(\xi) < \tau'$ for all $\xi < \nu$. $\square$

Remark: From lemma 30 one can prove that $\varkappa_\Theta$ is recursively regular. The following result is also true: If τ is recursively regular, and τ is projectible to ω (i.e. there is a one-one mapping π from τ to ω which is τ-recursive in constants less than τ) then there is a normal computation theory $(\Theta, ||_\Theta)$ on ω such that $\varkappa_\Theta = \tau$.

Lemma 31: If $\mathcal{L}$ is a normal list then $\kappa^{\mathcal{L}}$ is the least $\mathcal{L}$-regular ordinal.

Proof: Obviously all Θ-regular ordinals are limit ordinals $> \omega$ for any normal Θ. Let τ be a limit ordinal such that $\omega < \tau < \kappa^{\mathcal{L}}$. Since $\tau < \kappa^{\mathcal{L}}$ the induction which generates the convergent $\mathcal{L}$-computations does not stop at the stage τ. Hence there are $\mathcal{L}$-computations with length τ. Since τ is a limit ordinal such computations are applications of a functional to a function $\lambda x\{e\}^{\mathcal{L}}(x,a)$, where $\lambda x\{e\}^{\mathcal{L}}(x,a)$ is total, and $\tau = \sup\{|\{e\}^{\mathcal{L}}(x,a)| + 1 : x \in \omega\}$. Let π be defined by: $\pi(n) = |\{e\}^{\mathcal{L}}(n,a)|^{\mathcal{L}}$. Then $\tau = \sup\{\pi(n) : n \in \omega\}$. If τ was $\mathcal{L}$-regular then π would be $\mathcal{L}_{\tau}$-computable. By lemma 30 $\sup\{\pi(n) : n \in \omega\} < \tau$, a contradiction. Hence τ is not $\mathcal{L}$-regular. □

Lemma 32: If $(\Theta, ||_{\Theta})$ is a normal computation theory on ω, and $\mathcal{L}$ is a normal Θ-computable list, then there is a Θ-computable normal function π which has no Θ-regular fixed points less than $\kappa^{\mathcal{L}}$.

Proof: Let $(\Theta, ||_{\Theta})$ and $\mathcal{L}$ be as in the hypothesis. Since $\mathcal{L}$ is Θ-computable there is an index t such that for all x: $x \in C^{\mathcal{L}} \iff \langle t,x\rangle \in C_{\Theta}$. ($C^{\mathcal{L}}$ is the set of convergent $\mathcal{L}$-computations, C_{Θ} is the set of convergent Θ-computations.) If $\nu < \kappa_{\Theta}$ then the set $\{x : |x|^{\mathcal{L}} < \nu\}$ is Θ-computable. There is an ordinal $\mu < \kappa_{\Theta}$ such that for all $x : |x|^{\mathcal{L}} < \nu \Rightarrow |\langle t,x\rangle|_{\Theta} < \mu$. For the function π defined by: $\pi(|x|^{\mathcal{L}}) \simeq |\langle t,x\rangle|_{\Theta}$ is Θ-computable. By lemma 30 $\sup\{\pi(\eta) : \eta < \nu\} < \kappa_{\Theta}$. In the same way one can prove that if $\nu < \kappa_{\Theta}$ then there is an ordinal $\mu < \kappa_{\Theta}$ such that for all x: $|\langle t,x\rangle|_{\Theta} < \nu \Rightarrow |x|^{\mathcal{L}} < \mu$.

Let π be defined as follows: $\pi(0) = 0$, π is continuous at limits. If ν is $\nu' + 1$, let $\pi(\nu)$ be the least ordinal μ such that:

i) $\pi(\nu') < \mu$

ii) for all x: $|x|^{\mathcal{L}} < \nu \Rightarrow |\langle t,x\rangle|_\Theta < \mu$,

iii) for all x: $|\langle t,x\rangle|_\Theta < \nu \Rightarrow |x|^{\mathcal{L}} < \mu$.

By the discussion above $\pi(\nu)$ is defined, and $\pi(\nu) < \varkappa_\Theta$ when $\nu < \varkappa_\Theta$. By the construction π is normal. By an application of the second recursion theorem one can prove that π is Θ-computable.

It remains to prove that π has no Θ-regular fixed points less than $\varkappa^{\mathcal{L}}$. Suppose $\tau < \varkappa^{\mathcal{L}}$, lim τ and $\pi(\tau) = \tau$. Since $\tau < \varkappa^{\mathcal{L}}$ and lim τ there is an index e and a list a such that $\lambda x\{e\}^{\mathcal{L}}(x,a)$ is total, $|\{e\}^{\mathcal{L}}(x,a)|^{\mathcal{L}} < \tau$ for all x, and $\sup\{|\{e\}^{\mathcal{L}}(x,a)|^{\mathcal{L}} : x \in \omega\} = \tau$. By ii) $|\langle t,\langle e,x,a\rangle\rangle|_\Theta < \pi(\tau)$ for all x, hence $|\langle t,\langle e,x,a\rangle\rangle|_\Theta < \tau$ for all x. $\sup\{|\langle t,\langle e,x,a\rangle\rangle|_\Theta : x \in \omega\} = \tau$, for suppose that this supremum was equal to $\tau' < \tau$. Then by iii) $|\{e\}^{\mathcal{L}}(x,a)|^{\mathcal{L}} < \pi(\tau'+1)$ for all x. Now $\pi(\tau'+1) < \pi(\tau) = \tau$. So this is contrary to the fact that $\sup\{|\{e\}^{\mathcal{L}}(x,a)|^{\mathcal{L}} : x \in \omega\} = \tau$. If τ was Θ-regular then the function ρ defined by: $\rho(x) = |\langle t,\langle e,x,a\rangle\rangle|_\Theta$ for $x \in \omega$, would be Θ_τ-computable. By lemma 30 $\sup\{\rho(x) : x \in \omega\} < \tau$, contrary to what is proved above. So τ is not Θ-regular. □

In the proof of the next lemma we need the ordinal function ρ defined below.

Let $(\Theta, ||_\Theta)$ be a normal computation theory on ω. Since it is a computation theory the operations mentioned in the beginning of §6, such as substitution, primitive recursion etc. are allowed. Let us regard substitution: There is a Θ-computable mapping $g_1(e,f,n)$ such that for all e, f, a, x: $\{g_1(e,f,n)\}_\Theta(a) \simeq x \Leftrightarrow \exists u[\{e\}_\Theta(a) \simeq u$ and $\{f\}_\Theta(u,a) \simeq x]$, and if $\{g_1(e,f,n)\}_\Theta(a) \simeq x$ then for some u such that $\{e\}_\Theta(a) \simeq u$ and $\{f\}_\Theta(u,a) \simeq x$: $|\{e\}_\Theta(a)|$, $|\{f\}_\Theta(u,a)| < |\{g_1(e,f,n)\}_\Theta(a)|$. This operation has a finitary character, i.e. there are only finitely many natural predecessors of the computation $\{g_1(e,f,n)\}_\Theta(a)$, namely $\{e\}_\Theta(a)$ and $\{f\}_\Theta(u,a)$. The other operations mentioned in § 6 are also finitary.

Since $(\Theta, ||_\Theta)$ is normal there is a partial Θ-computable function p such that $p(x,y) \simeq 0$ if $x \in C_\Theta$ and $|x|_\Theta \leq |y|_\Theta$. This can also be regarded as a finitary operation. There is one natural predecessor of $\{\hat{p}\}_\Theta(x,y)$, namely x if $x \in C_\Theta$ and $|x|_\Theta \leq |y|_\Theta$, y if $|x|_\Theta > |y|_\Theta$. ($\hat{p}$ is a Θ-index for p).

Since ω is strongly Θ-finite the functional E_ω is weakly Θ-finite, where E_ω is defined by: $E_\omega(\varphi) \simeq 0$ if $\exists x\, \varphi(x) \simeq 0$, $E_\omega(\varphi) \simeq 1$ if $\forall x\, \exists y \neq 0\;\; \varphi(x) \simeq y$. There is a Θ-computable mapping $g_2(n)$ such that $\{g_2(n)\}_\Theta(e,a) \simeq E_\omega(\lambda x\{e\}_\Theta(x,a))$ for all e,a (the list a has length n). If $\{g(n)\}_\Theta(e,a) \simeq 0$ then the natural predecessors of $\{g(n)\}_\Theta(e,a)$ are those computations $\{e\}_\Theta(x,a)$ such that $\{e\}_\Theta(x,a) \simeq 0$ and $\{e\}_\Theta(x,a)$ has minimal length. If $\{g(n)\}_\Theta(e,a) \simeq 1$ then the natural predecessors of $\{g(n)\}_\Theta(e,a)$ are $\{e\}_\Theta(x,a)$ for all x. So this operation is not finitary.

Let $\nu < \kappa_\Theta$. There is an ordinal $\mu < \kappa_\Theta$ such that the following is true:

Substitution: If for some u $\{e\}_\Theta(a) \simeq u$ and $\{f\}_\Theta(u,a) \simeq x$, and $|\{e\}_\Theta(a)|$, $|\{f\}_\Theta(u,a)| < \nu$ then $|\{g_1(e,f,u)\}_\Theta(a)| < \mu$.
p : If $|x|_\Theta < \nu$ or $|y|_\Theta < \nu$ then $|\{\hat{p}\}_\Theta(x,y)| < \mu$.
E_ω : If for some x $\{e\}_\Theta(x,a) \simeq 0$ and $|\{e\}_\Theta(x,a)| < \nu$ then $|\{g_2(n)\}_\Theta(e,a)| < \mu$. If for all x there is a $y \neq 0$ such that $\{e\}_\Theta(x,a) \simeq y$ and $|\{e\}_\Theta(x,a)| < \nu$ then $|\{g_2(n)\}_\Theta(e,a)| < \mu$.
In addition there are clauses for the other operations mentioned in § 6.

There is a $\mu < \kappa_\Theta$ which satisfies these conditions. Otherwise we would obtain a contradiction to lemma 30. Let $\rho(\nu)$ be the least ordinal μ which satisfies the conditions above. Then $\rho(\nu) < \kappa_\Theta$ when $\nu < \kappa_\Theta$, and ρ is Θ-computable.

ρ is not necessarily continuous at limit ordinals. If $(\Theta, ||_\Theta)$ is the computation theory obtained from a normal list $\mathcal{L}$ it can be proved that $\rho(\nu) <$ the least limit ordinal $> \nu$. If $E_\omega(\lambda x\{e\}^{\mathcal{L}}(x,a)) \simeq 0$ and $\sup\{|\{e\}^{\mathcal{L}}(x,a)|^{\mathcal{L}} + 1 : x \in \omega\} = \lambda$, and $\lim \lambda$, then $\sup\{\rho(\nu) : \nu < \lambda\} = \lambda < \rho(\lambda)$. Hence ρ is not continuous at λ.

Suppose $\tau < \kappa_\Theta$ is a limit ordinal such that $\nu < \tau \Rightarrow \rho(\nu) < \tau$. If $\rho(\tau) = \tau$ then $(\Theta_\tau, ||_{\Theta\tau})$ is a normal computation theory on ω. $\rho(\tau) = \tau$ iff the following is true: For all e,a if $|\{e\}_\Theta(x,a)| < \tau$ for all x, then $|\{g_2(n)\}_\Theta(e,a)| < \tau$.

<u>Lemma 33</u>: If $(\Theta, ||_\Theta)$ is a normal computation theory on ω, and π is a normal Θ-computable function, then there is a normal Θ-computable list $\mathcal{L}$ such that $\kappa^{\mathcal{L}}$ is Θ-regular and a fixed point for π.

Proof: There is an index e_0 such that $\{e_0\}_\Theta(c)\downarrow$ iff $\lambda x\{e\}_\Theta(x)$ is total, in which case $|\{e\}_\Theta(x,a)|_\Theta < |\{e_0\}_\Theta(e)|_\Theta$ for all x. If f is a Θ-computable total function let $\mathrm{Ord}(f) = \inf\{|\{e_0\}_\Theta(e)|_\Theta : f = \lambda x\{e\}_\Theta(x)\}$.

Let the functional F be defined as follows.
If f is Θ-computable and total, let $\nu = \mathrm{Ord}(f)$, and let $\mu = \sup(\pi(\nu), \rho(\nu))$. Let

$$F(\langle f,n,0\rangle) = \begin{cases} 0 & \text{if } n = \langle e,a,y\rangle,\ \{e\}_\Theta(a) \simeq y,\ |\{e\}_\Theta(a)|_\Theta < \nu \\ 1 & \text{otherwise} \end{cases}$$

$$F(\langle f,n,1\rangle) = \begin{cases} 0 & \text{if } n = \langle e,a,y\rangle,\ \{e\}_\Theta(a) \simeq y,\ |\{e\}_\Theta(a)|_\Theta < \mu \\ 1 & \text{otherwise} \end{cases}$$

If g is not Θ-computable, or g is not of the form $\langle f,n,0\rangle$ or $\langle f,n,1\rangle$, let $F(g) = 1$. Then F is weakly Θ-computable. Hence the list $\mathcal{L} = E,F$ is Θ-computable.

If f is recursive in $\mathcal{L}$ and total then f is Θ-computable, hence $\mathrm{Ord}(f)$ is defined. Let $\lambda = \sup\{\mathrm{Ord}(f) : f$ is total and recursive in $\mathcal{L}\}$.

Suppose $f = \lambda x\{e\}^{\mathcal{L}}(x)$ is total. Let $\nu = \mathrm{Ord}(f)$.

$\lambda n\ F(\langle f,n,0\rangle)$ is the characteristic function of the set $B_e = \{\langle e',a,y\rangle : \{e'\}_\Theta(a) \simeq y$ and $|\{e'\}_\Theta(a)|_\Theta < \nu\}$. The set $C_e = \{\langle e',a,y\rangle : \{e'\}_\Theta(a) \simeq y$ and $|\{e'\}_\Theta(a)|_\Theta < \mu\}$ has the characteristic function $\lambda n\ F(\langle f,n,1\rangle)$ $(\mu = \sup(\pi(\nu),\rho(\nu))$. Hence B_e and C_e are recursive in $\mathcal{L}$, and $\mathcal{L}$-indexes for the characteristic functions of B_e and C_e can be found uniformly from e.

Let $\hat{p}$ be the index for the function p mentioned in the construction of ρ. If $\inf(|x|_\Theta,|y|_\Theta) < \nu$ then $|\{\hat{p}\}_\Theta(x,y)|_\Theta < \rho(\nu)$ by the construction of ρ. Hence $|\{\hat{p}\}_\Theta(x,y)|_\Theta < \mu$. The set $\{(x,y) : |x|_\Theta < \nu$ or $|y|_\Theta < \nu$, and $\langle\hat{p},x,y,0\rangle \in C_e\}$ is a prewellordering of length ν. It is recursive in $\mathcal{L}$ since B_e and C_e are recursive in $\mathcal{L}$. By lemma 20 each prewellordering which is recursive in $\mathcal{L}$ has length $< \kappa^{\mathcal{L}}$. Hence $\nu < \kappa^{\mathcal{L}}$, and $\lambda \leq \kappa^{\mathcal{L}}$.

By uniformity there is a primitive recursive mapping g such that $\{g(e)\}^{\mathcal{L}}(x,y) \simeq 0$ if $|x|_\Theta < \nu$ and $|x|_\Theta \leq |y|_\Theta$, $\{g(e)\}^{\mathcal{L}}(x,y) \simeq 1$ if $|x|_\Theta > |y|_\Theta$ and $|y|_\Theta < \nu$.

There is an index e_1 such that $\{e_1\}^{\mathcal{L}}(e)\downarrow$ iff $\lambda x\{e\}^{\mathcal{L}}(x)$ is total, in which case $|\{e\}^{\mathcal{L}}(x)|^{\mathcal{L}} < |\{e_1\}^{\mathcal{L}}(e)|^{\mathcal{L}}$ for all x. If f is total and recursive in $\mathcal{L}$ let $\mathrm{Ord}^{\mathcal{L}}(f) = \inf\{|\{e_1\}^{\mathcal{L}}(e)|^{\mathcal{L}} : f = \lambda x\{e\}^{\mathcal{L}}(x)\}$.

If $\mu < \kappa^{\mathcal{L}}$ then by a simple diagonalization method one can construct a function f which is total and recursive in $\mathcal{L}$, and f is different from all functions f' with $\mathrm{Ord}^{\mathcal{L}}(f') < \mu$. Hence $\mathrm{Ord}^{\mathcal{L}}(f) \geq \mu$, and $\sup\{\mathrm{Ord}^{\mathcal{L}}(f) : f$ is recursive in $\mathcal{L}\} = \kappa^{\mathcal{L}}$.

If $\mu < \kappa^{\mathcal{L}}$ let $\mathcal{F}_\mu = \{f : \mathrm{Ord}^{\mathcal{L}}(f) < \mu\}$. Let $\nu = \sup\{\mathrm{Ord}(f) : f \in \mathcal{F}_\mu\}$. Claim $\nu < \lambda$.

Proof: Let $D = \{e : |\{e_1\}^{\mathcal{L}}(e)| < \mu\}$. Then D is recursive in $\mathcal{L}$, and $f \in \mathcal{F}_\mu$ iff $f = \lambda x\{e\}^{\mathcal{L}}(x)$ for some $e \in D$. The set $\bigcup_{e\in D} B_e$ is recursive in $\mathcal{L}$ since B_e is recursive in $\mathcal{L}$ uniformly in e.

$\bigcup_{e\in D} B_e = \{\langle e',a,y\rangle : \{e'\}_\Theta(a) \simeq y$ and $|\{e'\}_\Theta(a)|_\Theta < \nu\}$. If f is a total Θ-computable function such that $Ord(f) \leq \nu$ then there is an index e such that $f = \lambda x\{e\}_\Theta(x)$, and $\forall x\, \exists y \langle e,x,y\rangle \in \bigcup_{e\in D} B_e$. Hence the set $E = \{e : \forall x\, \exists y\langle e,x,y\rangle \in \bigcup_{e\in D} B_e\}$ contains Θ-indexes for all total Θ-computable functions f with $Ord(f) \leq \nu$, in particular E contains Θ-indexes for all $f \in \mathcal{F}_\mu$. E is recursive in $\mathcal{L}$. By a diagonalization method one can construct a function f' which is recursive in $\mathcal{L}$, and such that f' is different from all total Θ-computable functions with Θ-index in E. Hence $Ord(f') > \nu$, and $\nu < \lambda$.

Let the ordinal function σ be defined by: If $\nu < \lambda$ let $\sigma(\nu) = \inf\{Ord^{\mathcal{L}}(f) : Ord(f) > \nu\}$. Then $\sup\{\sigma(\nu) : \nu < \lambda\} = \varkappa^{\mathcal{L}}$, for suppose $\sup\{\sigma(\nu) : \nu < \lambda\} = \mu < \varkappa^{\mathcal{L}}$. Then for all $\nu < \lambda$ there is an f such that $Ord(f) > \nu$ and $Ord^{\mathcal{L}}(f) < \mu$. Hence $\sup\{Ord(f) : Ord^{\mathcal{L}}(f) < \mu\} = \lambda$, contrary to the claim.

To prove that σ is recursive in $\mathcal{L}$ it is enough to prove that R is recursively enumerable in $\mathcal{L}$, where $R = \{(x,y) : x,y \in C^{\mathcal{L}}$ and $\sigma(|x|^{\mathcal{L}}) = |y|^{\mathcal{L}}\}$. $(x,y) \in R \iff x,y \in C^{\mathcal{L}}$ and $\exists e[\lambda t\{e\}^{\mathcal{L}}(t)$ is total and $|x|^{\mathcal{L}} < Ord(f)$ and $|y|^{\mathcal{L}} = Ord^{\mathcal{L}}(f)$ and $\forall e'(Ord^{\mathcal{L}}(f') < Ord^{\mathcal{L}}(f) \Rightarrow Ord(f') \leq |x|^{\mathcal{L}})]$ (where $f = \lambda t\{e\}^{\mathcal{L}}(t)$, $f' = \lambda x\{e'\}^{\mathcal{L}}(x)$.) "$\lambda t\{e\}^{\mathcal{L}}(t)$ is total" can be expressed by "$\{e_1\}^{\mathcal{L}}(e)\downarrow$". "$|y|^{\mathcal{L}} = Ord^{\mathcal{L}}(f)$" can be expressed by "$|y|^{\mathcal{L}} = |\{e_1\}^{\mathcal{L}}(e)|^{\mathcal{L}}$", which is recursively enumerable in $\mathcal{L}$. "$Ord^{\mathcal{L}}(f') < Ord^{\mathcal{L}}(f)$" can be expressed by "$|\{e_1\}^{\mathcal{L}}(e')|^{\mathcal{L}} < |\{e_1\}^{\mathcal{L}}(e)|^{\mathcal{L}}$". It remains to find expressions for "$|x|^{\mathcal{L}} < Ord(f)$" and "$Ord(f') \leq |x|^{\mathcal{L}}$". This is done in the next section.

Let $f = \lambda x\{e\}^{\mathcal{L}}(x)$ be total, and $Ord(f) = \nu$. Then there is an index $\hat{e}$ such that $f = \lambda x\{\hat{e}\}_\Theta(x)$, $|\{\hat{e}\}_\Theta(x)|_\Theta < \nu$ for all x, and $\nu = |\{e_0\}_\Theta(\hat{e})|_\Theta$. Also $\forall x\, \exists y(\langle \hat{e},x,y\rangle \in B_e)$. We can find such an index $\hat{e}$ by asking questions about B_e and C_e, which are recursive in $\mathcal{L}$, uniformly in e. There is a function t which is partial recursive in $\mathcal{L}$ such that if $\lambda x\{e\}^{\mathcal{L}}(x)$ is total then $t(e)$ is defined, and $t(e)$ is such a Θ-index for f. An $\mathcal{L}$-index for t can be constructed from the primitive recursive mapping g. There is also a primitive recursive mapping h such that if $f = \lambda x\{e\}^{\mathcal{L}}(x)$ is total and $\mathrm{Ord}(f) = \nu$ then

$\{h(e)\}^{\mathcal{L}}(x,y) \simeq 0$ if $|x|_{\Theta} \leq \nu$ and $|x|_{\Theta} \leq |y|^{\mathcal{L}}$,

$\{h(e)\}^{\mathcal{L}}(x,y) \simeq 1$ if $|x|_{\Theta} \leq \nu$ and $|x|_{\Theta} > |y|^{\mathcal{L}}$.

"$|x|^{\mathcal{L}} < \mathrm{Ord}(f)$" can be expressed by "$\{h(e)\}^{\mathcal{L}}(\langle e_0, e\rangle, x) \simeq 1$",

"$\mathrm{Ord}(f') \leq |x|^{\mathcal{L}}$" can be expressed by "$\{h(e)\}^{\mathcal{L}}(\langle e_0, e'\rangle, x) \simeq 0$".

This proves that R is recursively enumerable in $\mathcal{L}$, and hence that σ is $\mathcal{L}$-recursive.

As mentioned before $\sup\{\sigma(\nu) : \nu < \lambda\} = \kappa^{\mathcal{L}}$, and $\lambda \leq \kappa^{\mathcal{L}}$. It follows from lemma 30 that $\lambda = \kappa^{\mathcal{L}}$.

Next we prove that if $\nu < \kappa^{\mathcal{L}}$ then $\pi(\nu) < \kappa^{\mathcal{L}}$ and $\rho(\nu) < \kappa^{\mathcal{L}}$. Let $\nu < \kappa^{\mathcal{L}}$. Choose e such that $f = \lambda x\{e\}^{\mathcal{L}}(x)$ is total, and $\mathrm{Ord}^{\mathcal{L}}(f) = \nu' \geq \nu$. Let $\mu = \sup(\pi(\nu'), \rho(\nu'))$. C_e is recursive in $\mathcal{L}$. From C_e one can construct a function f' which is total and recursive in $\mathcal{L}$, and f' is different from all functions f'' with $\mathrm{Ord}(f'') \leq \mu$. Hence $\mu < \mathrm{Ord}(f')$. Since $\mathrm{Ord}(f') < \lambda = \kappa^{\mathcal{L}}$, $\mu < \kappa^{\mathcal{L}}$. Hence $\rho(\nu') < \kappa^{\mathcal{L}}$ and $\pi(\nu') < \kappa^{\mathcal{L}}$, and $\rho(\nu) < \kappa^{\mathcal{L}}$ and $\pi(\nu) < \kappa^{\mathcal{L}}$ since $\nu \leq \nu'$. This proves that $\kappa^{\mathcal{L}}$ is a fixed point for π

To prove that $\kappa^{\mathcal{L}}$ is Θ-regular it is enough to prove that if $|\{e\}_{\Theta}(x,a)|_{\Theta} < \kappa^{\mathcal{L}}$ for all x, then there is an ordinal $\mu < \kappa^{\mathcal{L}}$ such $|\{e\}_{\Theta}(x,a)|_{\Theta} < \mu$ for all x. So suppose $|\{e\}_{\Theta}(x,a)|_{\Theta} < \kappa^{\mathcal{L}}$ for all x. From the functions defined earlier in this proof one can construct an index e' such that $\{e\}_{\Theta}(x,a) \simeq \{e'\}^{\mathcal{L}}(x,a)$ for all x, and $|\{e\}_{\Theta}(x,a)|_{\Theta} < |\{e'\}^{\mathcal{L}}(x,a)|^{\mathcal{L}}$ for all x. There is a $\mu < \kappa^{\mathcal{L}}$ such that $|\{e'\}^{\mathcal{L}}(x,a)|^{\mathcal{L}} < \mu$ for all x. (Let μ be the length of the computation $E(\lambda x\{e'\}^{\mathcal{L}}(x,a))$.) Hence $|\{e\}_{\Theta}(x,a)|_{\Theta} < \mu$ for all x.

□

<u>Theorem 11</u>: Let $(\Theta, ||_\Theta)$ be a normal computation theory on ω. Then Θ is Mahlo iff Θ has property 1.

Proof: Suppose Θ is Mahlo. Let $\mathcal{L}$ be a normal Θ-computable list. By lemma 32 there is a normal Θ-computable ordinal function π which has no Θ-regular fixed points less than $\kappa^{\mathcal{L}}$. Since Θ is Mahlo there is a Θ-regular fixed point for π less than κ_Θ. Hence $\kappa^{\mathcal{L}} < \kappa_\Theta$, and Θ has property 1.

Suppose Θ has property 1. Let π be a normal Θ-computable ordinal function. By lemma 33 there is a normal Θ-computable list $\mathcal{L}$ such that $\kappa^{\mathcal{L}}$ is Θ-regular and a fixed point for π. Since Θ has property 1 $\kappa^{\mathcal{L}} < \kappa_\Theta$. Hence π has a Θ-regular fixed point less than κ_Θ. Hence Θ is Mahlo. $\square$

§ 10 CALCULATION OF THE LENGTHS OF SOME COMPUTATIONS

In this chapter we will prove some lemmas and introduce some notions which will be useful in § 11. Let $\mathfrak{A}$ be a computation domain with subindividuals S and individuals I, and let $\mathcal{L}$ be a normal list containing the functional E. We write $\{e\}$ instead of $\{e\}^{\mathcal{L}}$. Many of the proofs in this chapter are rather technical with few new ideas, so they will be omitted. However, a few of the proofs are given as examples.

<u>Lemma 34</u>: For each m, n there is a function S^n_m such that for all e, $x_1 \ldots x_n \in N$, all $y_1 \ldots y_m \in I$: $\{e\}(x_1 \ldots x_n, y_1 \ldots y_m) \simeq \{S^n_m(e, x_1 \ldots x_n)\}(y_1 \ldots y_m)$, and if $\{e\}(x_1 \ldots x_n, y_n \ldots y_m)\downarrow$ then $|\{S^n_m(e, x_1 \ldots x_n)\}(y_1 \ldots y_m)| = |\{e\}(x_1 \ldots x_n, y_1 \ldots y_m)| + n$.

<u>Remark</u>: The functions S^n_m are primitive recursive. There is a primitive recursive function f such that $f(n,m)$ is an index for S^n_m, i.e. $S^n_m = \{f(n,m)\}_{PR}$. If g is a primitive recursive function with index e (i.e. $g = \{e\}_{PR}$) then $|\{e\}(a)| < \omega$ for all a.

<u>Lemma 35</u>: There is a primitive recursive function f and a function $g_1 : \omega^2 \to \omega$ such that for all $e \in N$, all lists a of length n, all lists b with length m: $\{e\}(a) \simeq \{f(e,m)\}(a,b)$; and if $|\{e\}(a)| = \lambda + k$ where $\lambda = 0$ or $\lim \lambda$, $k < \omega$ then $|\{f(e,m)\}(a,b)| < \lambda + g_1(k,n)$.

<u>Lemma 36</u> (Second recursion theorem): There is a function $g_2 : \omega^2 \to \omega$ with the following property: For all $t \in N$ there is an e such that $\{t\}(e,a) \simeq \{e\}(a)$ for all a. If $|\{t\}(e,a)| \leq \lambda + k$ where λ, k are as in lemma 35, then $|\{e\}(a)| < \lambda + g_2(k,n)$. ($n$ is the length of the list a.)

Definitions: Let $\mathcal{F}$ be a functional whose domain is a subset of $X \times I^n$ where X is the set of all partial functions $I^n \to S$. $\mathcal{F}$ is monotone if for all $\varphi, \psi \in X$, all $a \in I^n$: $\mathcal{F}(\varphi,a)\downarrow$ and $\varphi \subseteq \psi \to \mathcal{F}(\psi,a) \simeq \mathcal{F}(\varphi,a)$.

$\mathcal{F}$ is weakly partial recursive in $\mathcal{L}$ if there is a primitive recursive function f such that for all $e \in N$, all $a \in I^n$, all $b \in I^m$: $\mathcal{F}(\lambda x\ \{e\}(x,b),a) \simeq \{f(m)\}(e,a,b)$.

Lemma 37 (First recursion theorem): There is a function $g_3 : \omega^2 \to \omega$ with the following property:

Let $\mathcal{F}$ be a monotone functional which is weakly partial recursive in $\mathcal{L}$. Then there is a least φ such that for all $a \in I^n$: $\mathcal{F}(\varphi,a) \simeq \varphi(a)$, and this φ is partial recursive in $\mathcal{L}$. There is an index $\hat{\varphi}$ for φ such that if $|\{f(0)\}(\hat{\varphi},a)| = \lambda + k$, λ and k as in lemma 35, then $|\{\hat{\varphi}\}(a)| < \lambda + g_3(k,n)$.

Proof: Let f be the primitive recursive function which proves that $\mathcal{F}$ is weakly partial recursive in $\mathcal{L}$. Let $\hat{\varphi} = S_n^2(i,f(0),j)$, where $\{i\}(e_1,e_2,a) \simeq \{e_1\}(\{e_2\}_{PR}(0),a)$, and j is an index found by the recursion theorem for primitive recursive functions such that $\{j\}_{PR}(t) = S_n^2(i,f(0),j)$ for all t. In [16] it is proved that $\{\hat{\varphi}\}$ is the least solution to the equality: $\forall a \in I^n(\mathcal{F}(\varphi,a) \simeq \varphi(a))$. $\{\hat{\varphi}\}(a) \simeq \{S_n^2(i,f(0),j)\}(a) \simeq \{i\}(f(0),j,a) \simeq \{f(0)\}(\{j\}_{PR}(0),a) \simeq \{f(0)\}(S_n^2(i,f(0),j),a) \simeq \{f(0)\}(\hat{\varphi},a)$.

Suppose $|\{f(0)\}(\hat{\varphi},a)| = \lambda + k$. Then $|\{f(0)\}(\{j\}_{PR}(0),a)| = \lambda + k$. Suppose $|\{e_1\}(u,a)| = \lambda + k$, where $u = \{e_2\}_{PR}(0)$. Let $|\{e_2\}_{PR}(0)| = k'$. Then $k' < \omega$. By several applications of lemma 34 and lemma 35 and some permutations of the argument list we obtain an index i and a function $h' : \omega^3 \to \omega$ such that $\{i\}(e_1,e_2,a) \simeq \{e_1\}(\{e_2\}_{PR}(0),a)$, and $|\{i\}(e_1,e_2,a)| < \lambda + h'(k,k',n)$. Let $e_2 = j$, $e_1 = f(0)$. Then $|\{i\}(f(0),j,a)| < \lambda + h'(k,k',n)$. By lemma 34 $|\{S_n^2(i,f(0),j)\}(a)| =$

$|\{i\}(f(0),j,a)|+2$. Hence $|\{\hat{\Phi}\}(a)| < \lambda + h'(k,k',n)+2$. Let $g_3(k,n) = h'(k,k',n)+2$, where $k' = |\{j\}_{PR}(0)|$. □

Let $C^{\mathcal{L}} = \{\langle e,a\rangle : \{e\}^{\mathcal{L}}(a)\downarrow\}$.

Lemma 38: There is a function p and a function $g_4 : \omega^3 \to \omega$ such that the following is true:

p is partial recursive in $\mathcal{L}$, $p(x,y)\downarrow$ iff $x \in C$ or $y \in C$, if $x \in C$ and $|x| \leq |y|$ then $p(x,y) \simeq 0$, if $|x| > |y|$ then $p(x,y) \simeq 1$. There is an index $\hat{p}$ for p such that if $\inf\{|x|,|y|\} = \lambda + k$, where $\lambda = 0$ or $\lim \lambda$, $k < \omega$, then $|\{\hat{p}\}(x,y)| < \lambda + g_4(k,m,n)$, where m is the number of arguments in x, n is the number of arguments in y.

Lemma 39 (Selection of numbers): There is a primitive recursive function f such that if $\exists n \in N\{e\}(n,a)\downarrow$ then $\{f(e)\}(a) \in N$, and $\{e\}(\{f(e)\}(a),a)\downarrow$. If $\inf\{|\{e\}(n,a)| : n \in N\} = \lambda + k$ where $\lambda = 0$ or $\lim \lambda$, $k < \omega$ then $|\{f(e)\}(a)| < \lambda + \omega$.

Proof: In §4 (theorem 3), this lemma is proved in the following way: Let ψ be the partial function defined by:

$$\psi(r,n,e,a) \simeq \begin{cases} n & \text{if } |\{e\}(n,a)| \leq |\{r\}(n+1,e,a)| \\ \{r\}(n+1,e,a) & \text{if } |\{e\}(n,a)| > |\{r\}(n+1,e,a)| . \end{cases}$$

By the second recursion theorem there is an r such that $\psi(r,n,e,a) \simeq \{r\}(n,e,a)$ for all n,e,a. The selection function is $\lambda a\{r\}(0,e,a)$ for this r. Let $f(e)$ be an index for $\lambda a\{r\}(0,e,a)$. It can be proved that $\{r\}(0,e,a)$ = the least n such that $\{e\}(n,a)\downarrow$ and $|\{e\}(n,a)| \leq |\{r\}(n+1,e,a)|$. Such n exists iff $\exists n \in N\{e\}(n,a)\downarrow$.

To prove the ordinal inequality we construct an index $\hat{\psi}$ for ψ. $\hat{\psi}$ can be found such that if $\inf\{|\{e\}(n,a)|,|\{r\}(n+1,e,a)|\} = \lambda_1 + k_1$ then $|\{\hat{\psi}\}(r,n,e,a)| < \lambda_1 + h_1(k_1,m)$, where h_1 is a function $\omega^2 \to \omega$, m is the length of the list a. (h_1 can be constructed from the function g_4 in lemma 38.)

Suppose $\{r\}(0,e,a) \simeq n$, and $\inf\{|\{e\}(n,a)| : n \in N\} = \lambda + k$, where r is the index found by the second recursion theorem, as mentioned above. By induction on i we prove that there is a function $j_i : \omega^2 \to \omega$ such that $|\{r\}(i,e,a)| < \lambda + j_i(k,m)$, for all $i \leq n$. The induction goes as follows. Suppose j_i has the desired properties, $i > 0$. We want to construct j_{i-1}. $\inf\{|\{e\}(i-1,a)|, |\{r\}(i,e,a)|\} \leq |\{r\}(i,e,a)| < \lambda + j_i(k,m)$ by the induction hypothesis. Let $\lambda_1 = \lambda$, $k_1 = j_i(k,m)$. By the above $|\{\hat{\psi}\}(r,i-1,e,a)| < \lambda + h_1(k_1,m)$. By lemma 36 there is a function $g_2 : \omega^2 \to \omega$ such that $|\{r\}(i-1,e,a)| < \lambda + g_2(h_1(k_1,m),m+2)$. Let $j_{i-1}(k,m) = g_2(h_1(k_1,m),m+2)$.

Hence $|\{r\}(0,e,a)| < \lambda + j_o(k,m)$. Let $f(e) = S^2_m(r,0,e)$. Then $\{f(e)\}(a) \simeq \{r\}(0,e,a)$, and $|\{f(e)\}(a)| < \lambda + j_o(k,m) + 2 < \lambda + \omega$.

□

The notion of a recursion structure will be useful in the next paragraph. Recursion structures were originally studied in Bergstra [23] and independently in Moldestad-Normann [24]. At our present level of generality, the suitable definition of recursion structures will be somewhat different from the one in those papers, but the main properties are the same.

<u>Definition</u>: Let S be a set of subindividuals with pairing and depairing functions as in the previous paragraphs.

<u>a</u> By the type language τ we mean the first order language with equality and the following non-logical symbols:

Relation symbols: I_i (for type (i) over S) and $\underline{\omega}$ (for ω).

Function symbols: $\underline{E}_i : \underline{T}_{i+1} \times \underline{T}_i \to \underline{\omega}$ to be given the interpretation of the evaluation functions.

$\underline{p}$ for each primitive recursive function $p : \omega \to \omega$, symbols for pairing and depairing on S.

<u>b</u> By a S-type-structure we mean a sequence $M = \langle M_i \rangle_{i \in \omega}$ such that

i $\omega \subseteq M_0 \subseteq S$ and M_0 is closed under pairing and depairing.

ii For all $i \in \omega$, $M_{i+1} \subseteq {}^{(M_i)}\omega = \{f : M_i \to \omega\}$.

τ has a canonical interpretation in any S-type-structure.

c Let M be a type structure. We obtain $\tau(M)$ from τ be adding constant symbols for each element of M.

We will in the sequel be most interested in countable type structures, even if S is uncountable. On the other hand we often want our type-structures to contain information about certain elements of $I = {}^S\omega$. To solve this technical problem we will rather investigate pairs $\langle M,\rho\rangle$ of the following sort.

Definition: Let M be a type structure, $\rho : M_1 \to I$. We call ρ a decode for M if for all $x \in M_1$, $\rho(x) \mid M_0 = x$.

As is verified before, given $f: S \to \omega$ we may regard f as a coded infinite sequence $f = \langle f_i\rangle_{i\in\omega}$.
A function $g: S \to \omega$ is called an enumeration of a subset of S if for all $n \neq 0$ there exists exactly one $i \in S$ such that $g(i) = n$.

Definition:

a Let $f = \langle f_i\rangle_{i\in\omega} \in I$. We call f a code if
$\forall i, n, m \in \omega\ (n \neq m \Rightarrow \lambda j \in \omega\ f_i(\langle j,n\rangle) \neq \lambda j \in \omega\ f_i(\langle j,m\rangle))$.

b Let f be a code, g an enumeration of a subset of S closed under pairing and depairing. Define the type structure $M = M(f,g)$ by induction as follows.
M_0 is the set enumerated by g, x_n^0 is the unique s such that $g(s) = n+1$.
Assume $M_i = \{x_n^i : n \in \omega\}$ is defined.
Let x_n^{i+1} be defined by: $x_n^{i+1}(x_m^i) = f_i(\langle m,n\rangle)$.
Let $M_{i+1} = \{x_n^{i+1} : n \in \omega\}$.

We call (f,g) a code for M.

c Let (f,g) be a code for M, ρ a decoding for M. $h : \omega \times S \to \omega$ is a code for ρ if $\rho(x_n^1) = \lambda s \in S\, h(n,s)$. We then say that (f,g,h) codes $\langle M,\rho\rangle$.

Lemma 40: If M is a countable type structure with decoding ρ, there exists a code for $\langle M,\rho\rangle$.

Proof: First, enumerate each $M_i = \{x_i^n : n \in \omega\}$.

Define $g(s) = \begin{cases} 0 & \text{if } s \notin M_o \\ n+1 & \text{if } s = x_o^n \end{cases}$

Define $f_i(\langle n,m\rangle) = x_{i+1}^m(x_i^n)$

$f_i(s) = 0$ for $s \notin \omega$

Define $f = \langle f_i\rangle_{i\in\omega}$.

Define $h(\langle n,s\rangle) = \rho(x_1^n)(s)$.

It is easily checked that (f,g,h) is a code for $\langle M,\rho\rangle$. □

We cannot expect to be able to do recursion theory on an arbitrary type structure. The closure properties we need are given on the following definition.

Definition: Let M be a type structure. We call M a recursion structure if the following two conditions hold:

1. If there is a formula φ in $\tau(M)$ in two variables x and y such that

$$M \models \forall x \in \underline{T}_n\ \exists y \in \underline{\omega}\ \varphi(x,y), \text{ then}$$

$$M \models \exists z \in \underline{T}_{n+1}\ \forall x \in \underline{T}_n\ \varphi(x,\underline{E}_{n+1}(z,x))$$

Before giving condition 2 we remark that 1 guarantees that the pairing and depairing functions may be extended from M_o to any M_i, and that any element of M_i may be identified with some element in M_j for $j > i$.

2. Let $n < m$ and let φ be a formula in $\tau(M)$ with two free variables x and y such that

$$M \models \forall x \in \underline{T}_n \; \exists y \in \underline{T}_m \; \varphi(x,y).$$

Then

$$M \models \exists z \in \underline{T}_m \; \forall x \in \underline{T}_n \; \varphi(x,(z)_x),$$

where $(z)_x(u) = z(\langle x,u \rangle)$.

We say that a recursion structure M is absolute if well-foundedness is absolute with respect to M.

Lemma 41: Let E be the functional defined in §4. Then the relation

a "(f,g,h) is a code for an absolute recursion structure M with a decode ρ" is recursive in E by a computation shorter than ω.

b If (f,g,h) is a code for a recursion structure M with decode ρ, then $\rho"M_1$ is uniformly recursive in E,f,g,h by a computation shorter than ω.

c If f,g,h,M,ρ are as above, then any statement about M expressible in $\tau(M)$ can be decided by a computation in E,f,g,h with length less than ω.

Proof of a: The statement "f is a code" can be expressed as QR where Q is a list of quantifiers ranging over ω, and R is a primitive recursive relation. "g is an enumeration of a subset of S, closed under pairing and depairing" can be expressed as QR, where Q is a list of quantifiers ranging over S, and R is primitive recursive. "h is a code for ρ" can be expressed by: $\forall n \in \omega \; \forall s \in S \; [g(s) \neq 0 \Rightarrow h(n,s) = f_1(\langle g(s),n \rangle)]$. "$M$ is a type structure" can be expressed as QR, where the quantifiers in Q range over ω, and R is primitive recursive, with parameters f and g.

There are countably many formulas in $\tau(M)$, and they can be enumerated by a function which is primitive recursive in f,g. A quantifier ranging over M can be translated to a number quantifier since M is countable. The expression "$M \models \varphi$" where φ is a sentence in $\tau(M)$ can be expressed as QR where Q is a list of number quantifiers, and R is primitive recursive with parameters f,g. The number of quantifiers in Q depend on φ. The statement "M is a recursion structure" is Δ_1^1 over the natural numbers, with parameters f,g. Finally the statement "M is absolute" can be expressed as: For all formulas φ in $\tau(M)$ with two free variables, both of type i, if $M \models \neg\exists \{y_k^i\}_{k=o}^{\infty} \forall k\ \varphi(y_k^i, y_{k+1}^i)$, then $\neg\exists \{y_k^i\}_{k=o}^{\infty} \forall k\ M \models \varphi(y_k^i, y_{k+1}^i)$. ($\{y_k^i\}_{k=o}^{\infty}$ is the sequence $y_o^i, y_1^i, \ldots$.) Because of the last quentifier $\neg\exists \{y_k^i\}_{k=o}^{\infty}$, this experssion is π_1^1 over the natural numbers, with parameters f,g. Hence the statement in a is also Π_1^1 over the natural numbers, with parameters f,g,h. It can be expressed as $\forall\alpha\ QR$ where α is ranging over the reals, and Q is a list of quantifers ranging over ω. R is primitive recursive in f,g,h. There is an index e_1 and a number n_1 such that $\{e_1\}$ is the characteristic function of R, and for each list —— $|\{e_1\}(—)| < n_1$. Expressing the quantifiers with E we find an index e_2 such that $\{e_2\}$ is the characteristic function of the statement in a, and $|\{e_2\}(f,g,h)| < n_2$ for some fixed number n_2, independant of f,g,h.

Remark: To obtain $|\{e_1\}(—)| < n_1$ R must be chosen with some care. We replace the schema of primitive recursion by quantifiers ranging over the reals. We obtain an R which is built up from the relation $m = n$, and the terms $+1$, $\alpha(x)$. Since we have a normal list $\mathcal{L}$, the equality relation on S is initial, hence so is "$m = n$". □

The proofs of b and c will be omitted.

Now let M be a recursion structure. We define recursion in M by relativizing the schemata in § 2 to M, i.e., let $S_M = M_o$, $I_M =$

$S_M \cup M_1$.

The closure properties of a recursion structure ensures that any M-recursive functional defined on some M_i will be an element of M_{i+1}. In an absolute recursion structure M, the computation tuples will be $\tau(M)$-definable.

The functional E is defined by a τ-formula in the full S-type hierarchy. Relativizing this definition to a recursion structure M, we obtain $E_M \in M_3$. We will regard recursion in lists $\mathcal{L}_M$ of objects from M_2 and M_3, containing E_M. In the rest of this paragraph we let all recursion structures in question be absolute. When working with a recursion structure M we will assume that a list $\mathcal{L}_M$ is given, although $\mathcal{L}_M$ will not be mentioned explicitly. We write $\{e\}_M(b) \simeq n$ instead of $\{e\}^{\mathcal{L}_M}(b) \simeq n$, where b is a list of objects from I_M. $|\{e\}_M(b)|$, $|\langle e,b\rangle|_M$ both denote the length of the computation $\{e\}^{\mathcal{L}_M}(b)$.

<u>Definition</u>: Let M be a countable recursion structure with decode ρ, a a list from I, $\tau < \kappa^a$.

<u>$\langle M,\rho\rangle$ is a (τ,a)-structure</u> if $a \subseteq \rho'' I_M$ and there is an ordinal ν such that for all $e \in N$, all finite lists b from I_M:

$|\{e\}(\rho(b))| < \tau \iff |\{e\}_M(b)|_M < \nu$ and

$|\{e\}(\rho(b))| < \tau \implies \{e\}(\rho(b)) \simeq \{e\}_M(b)$.

<u>Remark</u>: Suppose M is a (τ,a)-structure. Let τ',ν' be the greatest limit ordinals $\leq \tau,\nu$ respectively. Then for all $e \in N$ and all finite lists b from I_M: $|\{e\}(\rho(b))| < \tau' \iff |\{e\}_M(b)| < \nu'$.

If $\hat{p}$ is the index for the function p which compares the lengths of computations (lemma 38) then

$\{\hat{p}\}_M(x,y) \simeq 0$ if $x,y \in I_M$, $|x|_M \leq |y|_M$ and $|x|_M < \nu'$

$\simeq 1$ if $x,y \in I_M$, $|x|_M > |y|_M$ and $|y|_M < \nu'$

Because of this the function $\pi : \nu' \to \tau'$ defined by $\pi(|\{e\}_M(b)|_M) = |\{e\}(\rho(b))|$ is welldefined and order preserving. It can also be proved that if $e, e' \in N$, b, b' are finite lists from I_M and $|\{e\}(\rho(b))|$, $|\{e'\}(\rho(b'))| < \tau'$, then $\{e\}_M(b)$ is a subcomputation of $\{e'\}_M(b')$ in M iff $\{e\}(\rho(b))$ is a subcomputation of $\{e'\}(\rho(b'))$.

Lemma 42: Let a be a finite list from I. There is a countable recursion structure M with decoding ρ and a list $a_1 \in I_M$ such that $\rho(a_1) = a$ and M is a (κ^a, a)-structure.

Proof. Let $M^1 = \langle \omega, S, T_k \rangle_{k \in \omega}$. M^1 is a recursion structure. By the Skolem-Löwenheim theorem, let M^2 be a countable elementary submodel of M^1 with respect to $\tau(M^2)$, such that a is in I_{M^2}. We will obtain M from M^2 by a variant of the Mostowski isomorphism: Let $M_0 = M_0^2$, ψ be the identity on M_0. Assume M_i is defined as $\psi'' M_i^2$. For $x \in M_{i+1}^2$ let $\psi(x)(y) = x(\psi^{-1}(y))$, $y \in M_i^2$. Since M satisfies extentionality, so will M^2, and ψ is an 1-1 and faithfull to the atomic formulas. We leave it as an exercise to prove that M is absolute. The decoding ρ will be ψ^{-1} on M_1.

Remark: In the rest of this paragraph and in the next we will not distinguish between a and $\rho(a)$, since it will always be clear if we regard an element of I_M or of I. Note that $\rho \restriction ({}^{\omega}\omega)_M$ is the identity since $\omega \subseteq S_M = M_0$.

Lemma 43: There is an index e such that if x is a code for a countable recursion structure M, and there is an ordinal τ such that M is not a τ-structure, then $\{e\}(x)\downarrow$, and $\tau < |\{e\}(x)| < \tau + \omega$ for the least such τ.

Proof: Let η be the greatest ordinal such that M is an η-structure, and let $\tau = \eta + 1$. Let ν be the unique ordinal such that for all $i \in N$, all finite lists b from I_M:

$|\{i\}(b)| < \eta \iff |\{i\}_M(b)|_M < \nu$, and $|\{i\}(b)| < \eta \Rightarrow \{i\}(b) \simeq \{i\}_M(b)$. Since M is not a τ-structure there are an index e_0 and a list b_0 from I_M such that $|\{e_0\}(b_0)| = \eta$, and $|\{e_0\}_M(b_0)|_M = \nu$. To see that $|\{e_0\}_M(b_0)|_M = \nu$ we regard the subcomputations of $\{e_0\}(b_0)$. If $\{e_0\}$ is not an application of a functional then the immediate subcomputations of $\{e_0\}(b_0)$ have input lists from I_M, and length $< \eta$. Hence these subcomputations converge in M below ν, and $|\{e_0\}_M(b_0)|_M \leq \nu$. $|\{e_0\}_M(b_0)|_M < \nu$ is impossible since it implies $|\{e_0\}(b_0)| < \eta$. Hence $|\{e_0\}_M(b_0)|_M = \nu$. If $\{e_0\}$ is an application of a functional, i.e. $\{e_0\}(b_0) \simeq F(\lambda t\{e_1\}(t,b_0))$, then $|\{e_1\}(t,b_0)| < \eta$ for all $t \in I$, hence for all $t \in I_M$. Hence $|\{e_1\}_M(t,b_0)|_M < \nu$ for all $t \in I_M$, hence $|\{e_0\}_M(b_0)|_M = \nu$.

Since M is not a τ-structure there is also an index e_2 and a finite list b_2 from I_M such that $|\{e_2\}_M(b_2)|_M = \nu$, and either $|\{e_2\}(b_2)| > \eta$, or $|\{e_2\}(b_2)| = \eta$ and $\{e_2\}(b_2) \simeq i$, $\{e_2\}_M(b_2) \simeq j$, $i \neq j$.

Such e_0, b_0, e_2, b_2 can be found by searching through I_M, which (via a code x for M) is the same as a search through N. Let $\{e\}(x)$ be a computation which performs such a search. $|\{e\}(x)|$ will be τ + a finite number. □

<u>Notation</u>: Let WO denote the set of prewellorderings with domain included in ω.

§ 11 GAPS

Part A. On the lengths and order types of a- gaps.

Let a be a finite list if individuals. An ordinal τ is said to be a-constructive if $\tau = |\{e\}(a')|$, where $\{e\}(a')\downarrow$, and the list a' contains only natural numbers and elements from a . If τ is a-constructive then $\tau+1$, $\tau+\omega$, $\tau+\omega\cdot\omega$, $\tau+\omega^{\omega}$, ... are a-constructive. The set of a-constructive ordinals is not transitive, because it is countable, and it contains ordinals which are not countable. Hence the a-constructive ordinals lies in blocks. The a-gaps are the intervals between these blocks. If ν, ν' are ordinals let $[\nu,\nu')$ denote the set $\{\mu : \nu \leq \mu < \nu'\}$. $[\nu,\nu')$ is <u>a-gap</u> if $[\nu,\nu')$ contains no a-constructive ordinals, ν is a limit ordinal such that the a-constructive ordinals are cofinal in ν, and ν' is a-constructive. By convention we let $[\varkappa^{a},\infty)$ be an a-gap. To each a-gap $[\nu,\nu')$ (except $[\varkappa^{a},\infty)$) we can assign the a-constructive ordinal ν', hence there are at most countably many a-gaps. If τ is a-constructive then there are a-gaps between τ and $\varkappa^{a}$, hence the number of a-gaps is not finite. Since there are uncountable a-constructive ordinals there are a-gaps below $\varkappa^{a}$ of uncountable length. In this chapter we will calculate the lengths of some a-gaps, and the number of a-gaps below τ for some ordinals τ. An important tool for doing this is the countable recursion structures.

<u>Lemma 44</u>: Suppose τ is a-constructive, and $w \in WO$ is recursive in a . Let $\beta = |w|$. Then the order type of the a-gaps with length $\geq \tau$ is at least β .

Proof: To obtain a contradiction suppose that there are less than β a-gaps of length $\geq \tau$. Choose $i \in fld(w)$ such that $|i|_{w}$ is the number of a-gaps with length $\geq \tau$. There is an index e such that $\{e\}(j,a) \simeq 0$ for all $j <_{w} i$, and $|\{e\}(j,a)| \geq$ the first ordinal above a-gap number $|j|_W$ with length $\geq \tau$. Let μ be the length of the computation $\forall j <_W i \ (\{e\}(j,a) \simeq 0)$. Then μ is above all a-gaps with length $\geq \tau$, and μ is a-constructive. $\kappa^a < \mu + \aleph_1 + (\tau + \aleph_1) \cdot \aleph_1$, because $\mu + \aleph_1$ is greater than the first ordinal in the first a-gap above μ, $\tau + \aleph_1$ is greater than the length of any a-gap above μ + the first block of a-constructive ordinals above that a-gap, and the number of a-gaps above μ is less than $\aleph_1$. But the ordinal $\mu + \aleph_1 + (\tau + \aleph_1) \cdot \aleph_1$ is a-constructive since μ, τ and $\aleph_1$ are. Hence there is an a-constructive ordinal above κ^a, a contradiction. □

<u>Remark</u>: To prove that $\aleph_1$ is a-constructive for any a: There is an index e such that $\{e\}(w)\downarrow$ for all $w \in WO$, and $|\{e\}(w)| = |w|$ + a finite number. Hence $|\{e\}(w)| < \aleph_1$, and $\sup\{|\{e\}(w)| : w \in WO\} = \aleph_1$. The computation $\forall w \in WO \ \{e\}(w)\downarrow$ has length $\aleph_1$.

<u>Corollary</u>: If τ is a-constructive then there are λ^a a-gaps with length $\geq \tau$ below κ^a.

Proof: Since λ^a is the order type of the a-constructive ordinals, there are at most λ^a a-gaps below κ^a. Since $\lambda^a = \sup\{|w| : w \in WO$ and w is recursive in $a\}$, it follows from lemma 44 that there are at least λ^a a-gaps with length $\geq \tau$.

<u>Definition</u>: Let X be a subset of I, α an ordinal. <u>X is recursive in a below α</u> if there is an index e such that $\lambda x \{e\}(x,a)$ is the characteristic function of X and $\sup\{|\{e\}(x,a)| : x \in I\} < \alpha$. An ordinal <u>$\tau$ is recursive in a below α</u> if there is a prewellordering with length τ which is recursive in a below α.

<u>Theorem 12</u>: The length of an a-gap is a limit ordinal.

Proof: To obtain a contradiction suppose $[\nu, \lambda + m)$ is an a-gap, where λ is a limit ordinal, m is a finite number > 0. There are e, a'

such that $|\{e\}(a')| = \lambda + m$, where a' contains natural numbers and elements from the list a. We construct a sequence $B_o, B_1, \ldots, B_{m+1}$ of sets of individuals such that

(i) $B_o = \{\langle e, a'\rangle\}$,

(ii) each x in B_{k+1} is a subcomputation of a computation in B_k, $k = 0, \ldots, m$,

(iii) $\sup\{|x| : x \in B_k\} \geq \lambda$, $k = 0, \ldots, m+1$,

(iv) B_k is recursive in a below λ, $k = 0, \ldots, m+1$.

The construction goes as follows: Suppose B_k is constructed and satisfies (i) - (iv). B_k can be divided into 17 sets $C^1, \ldots, C^{17}$, where C^i contains those x in B_k which are convergent by the i-th clause of the operator Γ defined in § 2. Since B_k satisfies (iii) there is an i such that $\sup\{|x| : x \in C^i\} \geq \lambda$. B_{k+1} is constructed from C^i. We regard only the case $i = 10$ (the clause of substitution). The other cases are simpler.

$C^{10} = \{x : x \in B_k, x = \langle\langle 10, n, e_1, e_2\rangle, \text{—}\rangle\}$.

Case I: $\sup\{|\{e_1\}(\text{—})| : \exists n, e_2 \ \langle\langle 10, n, e_1, e_2\rangle, \text{—}\rangle \in B_k\} \geq \lambda$.

Let $B_{k+1} = \{\langle e_1, \text{—}\rangle : \exists n, e_2 \ \langle\langle 10, n, e_1, e_2\rangle, \text{—}\rangle \in B_k\}$.

Case II: $\sup\{|\{e_1\}(\text{—})| : \exists n, e_2 \ \langle\langle 10, n, e_1, e_2\rangle, \text{—}\rangle \in B_k\} < \lambda$.

Let $B_{k+1} = \{\langle e_2, \{e_1\}(\text{—}), \text{—}\rangle : \exists n \ \langle\langle 10, n, e_1, e_2\rangle, \text{—}\rangle \in B_k\}$.

In both cases B_{k+1} satisfies (i) - (iv). B_{k+1} satisfies (iv) because C^{10} is recursive in a below λ.

By (ii) each x in B_{m+1} is a convergent computation with length $< \lambda$. By (iii) $\sup\{ x| : x \in B_{m+1}\} = \lambda$. Let $\{e_1\}(a)$ be the following computation: Take all x in B_{m+1} and compute x. Then $\{e_1\}(a)\downarrow$, and by the above discussion $|\{e_1\}(a)| = \lambda$, contrary to the fact that λ is not a-constructive.

□

Definition: If α, β are ordinals, let $\underline{\gamma(\alpha,\beta)}$ be the first ordinal in the a-gap which is described by: it has length $\geq \alpha$, and there are β many a-gaps below it with length $\geq \alpha$.

Let M be a recursion structure, $\mathcal{L}_M$ a list, τ an ordinal, a a list of objects from I_M. τ is a-constructive in M if τ is a-constructive in the recursion theory built up from $\mathcal{L}_M$. a-gaps in this recursion theory are called a-gaps in M, and $\gamma_M(\alpha,\beta)$ is defined in the same way as $\gamma(\alpha,\beta)$.

Theorem 13: Let α be a limit ordinal, β a countable ordinal. Suppose α is recursive in a, and that there is a $w \in WO$ such that $|w| = \beta$ and w is recursive in a. If α, w are recursive in a below $\gamma(\alpha,\beta)$ then the a-gap starting with $\gamma(\alpha,\beta)$ has length α.

Proof: By lemma 44 $\gamma(\alpha,\beta)$ exists. It suffices to find an index e such that $|\{e\}(a)| = \gamma(\alpha,\beta) + \alpha$. We construct such an index. The computation $\{e\}(a)$ will say that $\gamma(\alpha,\beta)$ exists.

Let M be a countable recursion structure. Then either M is a $(\gamma(\alpha,\beta),a)$-structure, or M is not. In the first case there is a unique ordinal ν such that for all $i \in N$, all finite lists b from I_M: $|\{i\}(b)| < \gamma(\alpha,\beta) \iff |\{i\}_M(b)|_M < \nu$, $|\{i\}(b)| < \gamma(\alpha,\beta) \Rightarrow \{i\}(b) \simeq \{i\}_M(b)$.

The ordinal α_M is defined as follows. By assumption there is an index i such that $\lambda x\{i\}(x,a)$ is the characteristic function of a prewellordering with length α, and $\sup\{|\{i\}(x,a)| : x \in I\} < \gamma(\alpha,\beta)$. In the first case $\sup\{|\{i\}_M(x,a)|_M : x \in I_M\} < \nu$, and $\lambda x\{i\}_M(x,a)$ is the characteristic function of a prewellordering, x ranges over I_M. Let α_M be the length of this prewellordering. $\alpha_M \leq \alpha$ since I_M can be regarded as a subset of I.

In the first case the order type of the a-constructive ordinals below $\gamma(\alpha,\beta)$ is the same as the order type of the ordinals below ν which are a-constructive in M. The number of a-gaps is the same.

An a-gap below $\gamma(\alpha,\beta)$ has length $\geq \alpha$ iff the corresponding a-gap in M has length $\geq \alpha_M$. Hence there are β many a-gaps in M below ν with length $\geq \alpha_M$, so $\gamma_M(\alpha_M,\beta)$ exists, and $\nu \leq \gamma_M(\alpha_M,\beta)$. There are two possibilities: $\nu = \gamma_M(\alpha_M,\beta)$, or $\nu < \gamma_M(\alpha_M,\beta)$.

The idea of the proof is the following. Find indeces e_1, e_2, e_3 such that $|\{e_2\}(x,a)| < \gamma(\alpha,\beta)$ if the second case is true. If the first case is true then $\gamma(\alpha,\beta) \leq \inf\{|\{e_1\}(x,a)|, |\{e_2\}(x,a)|, |\{e_3\}(x,a)|\}$. If the first possibility is true then $|\{e_2\}(x,a)| = \gamma(\alpha,\beta)$. If the second possibility is true then $|\{e_3\}(x,a)| < \gamma(\alpha,\beta)+\alpha$. Here x is a code for M, i.e. $x = \langle f,g,h,n_1 \dots n_k\rangle$, where (f,g,h) is a code for M, and the numbers $n_1,\dots,n_k$ label the objects in the list $\mathcal{L}_M$. Given e_1, e_2, e_3 with the above properties let $\{e\}(a)$ be the following computation: $\forall x$ (x is the code for a countable recursion structure $\Rightarrow$ $\{e_1\}(x,a)\downarrow$ or $\{e_2\}(x,a)\downarrow$ or $\{e_3\}(x,a)\downarrow$). Then $\{e\}(a)\downarrow$. The length of the computation "$\{e_1\}(x,a)\downarrow$ or $\{e_2\}(x,a)\downarrow$ or $\{e_3\}(x,a)\downarrow$" is $\inf\{|\{e_1\}(x,a)|, |\{e_2\}(x,a)|, |\{e_3\}(x,a)|\}$ + a finite number. In any of the cases above this ordinal is less than $\gamma(\alpha,\beta)+\alpha$. By lemma 42 there is an x which brings us to the first case, first possibility. For this x the length is at least $\gamma(\alpha,\beta)$. To decide whether or not x is a code for a countable recursion structure can be done by a computation with finite length (lemma 41). Hence $\gamma(\alpha,\beta) < |\{e\}(a)| \leq \gamma(\alpha,\beta)+\alpha$. Hence $|\{e\}(a)| = \gamma(\alpha,\beta)+\alpha$ since the ordinals in the interval $[\gamma(\alpha,\beta), \gamma(\alpha,\beta)+\alpha)$ are not a-constructive ($\gamma(\alpha,\beta)$ is the first ordinal in an a-gap with length $\geq \alpha$).

Let e_1 be the index from lemma 43.

Construction of e_2. Let $\{e_2\}(x,a)$ be the following computation: For all $i \in N$, all lists b from I_M, if $|\{i\}_M(b)|_M < \gamma_M(\alpha_M,\beta)$, then compute $\{i\}(b)$. In the computation which decides if $|\{i\}(b)|_M < \gamma_M(\alpha_M,\beta)$, β is involved. β is recursive in a below $\gamma(\alpha,\beta)$, and facts about recursion in M can be decided by finite computations by lemma 41. Hence the computation is shorter than $\gamma(\alpha,\beta)$. In the first case $\{i\}(b)$ is a subcomputation of $\{e_2\}(x,a)$ for all i, b such that $|\{i\}(b)| < \gamma(\alpha,\beta)$. Hence $\gamma(\alpha,\beta) \leq |\{e_2\}(x,a)|$. If the first possibility is true all subcomputations $\{i\}(b)$ of $\{e_2\}(x,a)$ will be shorter than $\gamma(\alpha,\beta)$. Hence $|\{e_2\}(x,a)| = \gamma(\alpha,\beta)$.

Construction of e_3. Let $\{e_3\}(x,a)$ be the following computation: Search through N to find an i such that a), b), c) and d) are true.

a) $\{i\}_M(a)\downarrow$,

b) $\forall i' \in N[|\{i'\}_M(a)|_M < |\{i\}_M(a)|_M \Rightarrow \{i'\}(a)\downarrow]$,

c) the order type of the a-gaps in M below η with length $\geq \alpha_M$ is β, where $\eta = \sup\{|\{i'\}_M(a)|_M : |\{i'\}_M(a)| < |\{i\}_M(a)|_M\}$,

d) there is a u in the prewellordering with length α_M mentioned above such that $|\{i\}_M(a)|_M = \eta + |u|_M$, and $\sup\{|\{i'\}(a)| : |\{i'\}_M(a)|_M < |\{i\}_M(a)|_M\} + |u|$ is not a-constructive ($|u|_M$ is the ordinal of u in the prewellordering with length α_M, $|u|$ is the ordinal of u in the prewellordering with length α).

Suppose we are in the first case, second possibility. Then $\nu < \gamma_M(\alpha_M,\beta)$. Since ν is not the first ordinal in an a-gap in M with length $\geq \alpha_M$ there is an i such that $\nu \leq |\{i\}_M(a)|_M < \nu + \alpha_M$, and no ordinals between ν and $|\{i\}_M(a)|_M$ are a-constructive in M. Such an i satisfies a), b), c) and d). If $|\{i\}_M(a)|_M < \nu$ then either c) or d) will not be satisfied. In the first case, first possibility, no i satisfies a), b), c) and d).

As $\{e_3\}(x,a)$ diverges in the first case, first possibility, it remains only to compute $|\{e_3\}(x,a)|$ assuming the second possibility. By lemma 39 $|\{e_3\}(x,a)| = |\{e_4\}(i,x,a)|$ + a finite number, where i is chosen such that $|\{e_4\}(i,x,a)|$ is minimal, $\{e_4\}(i,x,a)$ is the computation for the conjunction of a), b), c) and d). If i satisfies a) - d) then $\nu \leq |\{i\}_M(a)|_M$ by the discussion above. Hence $\gamma(\alpha,\beta) \leq$

$|\{e_4\}(i,x,a)|$ because of the subcomputations for b). However $\gamma(\alpha,\beta)$ is an upper limit for the lengths of the subcomputations for b). The subcomputations for a) and c) are also bounded above by $\gamma(\alpha,\beta)$ as they only involve the prewellorderings with lengths α and β (recursive in a below $\gamma(\alpha,\beta)$), and facts about M. The computation for d) is first a search through I_M to find the u. This can be expressed as a search through N. To express that $\sup\{|\{i'\}(a)|:$ $|\{i'\}_M(a)|_M < |\{i\}_M(a)|_M\}+|u|$ is not a-constructive we get a computation with length $\gamma(\alpha,\beta) + |u| +$ a finite number. Now $|u| < \alpha$. This gives $|\{e_3\}(x,a)| < \gamma(\alpha,\beta)+\alpha$. □

Corollary 1:

a The first a-gap with length $\geq \omega, \omega+\omega, \omega+\omega+\omega,\ldots,\omega^2, \omega^2+\omega,\ldots$ $\ldots,\omega^n,\ldots,\omega^\omega,\ldots,\aleph_1,\aleph_1+\omega,\ldots,\aleph_1+\aleph_1,\ldots,\aleph_1^{\aleph_1},\ldots$ (ordinal exponentiation) has length equal to these ordinals respectively.

b a-gap number $1,2,\ldots,\omega,\omega+1,\ldots,\omega^2,\ldots,\omega^\omega,\ldots$ has length ω.

Corollary 2: If α is a limit ordinal which is recursive in a then there is an a-gap with length α.

Proof: There is an a-constructive ordinal $\tau < \kappa^a$ such that α is recursive in a below τ. Let β be the number of a-gaps with length $\geq \alpha$ below τ. There is a $w \in WO$ such that $\beta = |w|$, and w is recursive in a below the first a-gap above τ. By the theorem the first a-gap above τ with length $\geq \alpha$ has length α. □

Corollary 3: Suppose $\gamma(\alpha,\beta)$ exists, $\delta < \alpha$, $\lim \delta$, and δ is recursive in a below $\gamma(\alpha,\beta)$. Then the order type of the a-gaps below $\gamma(\alpha,\beta)$ with length $\geq \delta$ is $\lambda^a_{\gamma(\alpha,\beta)}$.
(Definition: λ^a_τ is the order type of the a-constructive ordinals less than τ for any ordinal τ.)

Proof: To obtain a contradiction suppose that the order type of the a-gaps below $\gamma(\alpha,\beta)$ with length $\geq \delta$ is ν, where $\nu < \lambda^a_{\gamma(\alpha,\beta)}$. By the definitions of $\gamma(\alpha,\beta)$ and $\gamma(\delta,\nu)$ $\gamma(\alpha,\beta) = \gamma(\delta,\nu)$. Since $\nu < \lambda^a_{\gamma(\alpha,\beta)}$ there is a $w \in WO$ such that $|w| = \nu$, and w is recursive in a below $\gamma(\alpha,\beta)$. By the theorem the a-gap starting with $\gamma(\delta,\nu)$ has length δ, contrary to the fact that this a-gap has length $\geq \alpha$. □

Corollary 4: The order type of the a-gaps below $\gamma(\omega+\omega,0)$ is $\gamma(\omega+\omega,0)$ (i.e. $\gamma(\omega,\gamma(\omega+\omega,0)) = \gamma(\omega+\omega,0)$).

Proof: By corollary 3 the order type of the a-gaps below $\gamma(\omega+\omega,0)$ is $\lambda^a_{\gamma(\omega+\omega,0)}$. So it is enough to prove that $\lambda^a_{\gamma(\omega+\omega,0)} = \gamma(\omega+\omega,0)$. This follows from the fact that all a-gaps below $\gamma(\omega+\omega,0)$ have length ω. If τ_1 and τ_2 are the first ordinals in two consecutive a-gaps below $\gamma(\omega+\omega,0)$ then the a-constructive ordinals between τ_1 and τ_2 have order type $\tau_2 - \tau_1$.

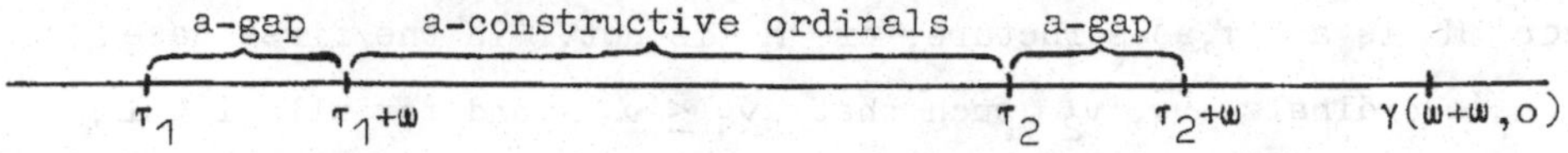

Hence $\lambda^a_\tau = \tau$ if τ is the first ordinal in any a-gap below $\gamma(\omega+\omega,0)$, and $\lambda^a_{\gamma(\omega+\omega,0)} = \gamma(\omega+\omega,0)$.

Part B. Reflection in a-gaps.

In [7] Harrington introduced a notion of reflection for recursion in normal objects of type ≥ 3, and he proves that some interesting ordinals are reflecting, In part B we prove that there are reflecting ordinals in many of the a-gaps.

Definition: An ordinal η is a-reflecting if for all $e \in N$: $\exists y|\{e\}(y,a)| < \eta \Rightarrow \exists y|\{e\}(y,a)| < \zeta$, where $\zeta = \sup\{\nu + 1 : \nu$ is a-constructive, $\nu < \eta\}$.

<u>Remark</u>: Any ordinal which is a-reflecting by the definition in [7] is also a-reflecting by this definition. The notion of reflection in [7] is reflection in the gap $[\varkappa^a, \infty)$. Here we will also consider reflection in the a-gaps below $\varkappa^a$.

If $[\zeta, \xi)$ is an a-gap then trivially ζ is a-reflecting. The interesting ordinals in this connection are the ordinals in the set $(\zeta, \xi] = \{\nu : \zeta < \nu \leq \xi\}$.

Suppose $[\zeta, \xi)$ is an a-gap. Let $P = \{n : |\{n\}(a)| < \zeta\}$. Let $\zeta^P = \sup\{|\{e\}(a,P)| : \exists\tau(\lim\tau \text{ and } |\{e\}(a,P)| < \tau < \xi)\}$.

<u>Theorem 14</u>: Suppose $[\zeta, \xi)$ is an a-gap. Then ζ^P is a-reflecting.

Proof: If the length of the gap is ω then $\zeta^P = \zeta$, and the theorem is trivial. So suppose that the length $> \omega$. Then $\zeta < \zeta^P$. Suppose $\exists y |\{e\}(y,a)| < \zeta^P$. Then there is an index e_0 and a limit ordinal τ such that $\exists y |\{e\}(y,a)| < |\{e_0\}(a,P)| < \tau < \xi$, $\zeta < \tau$.

Let M be a countable recursion structure. There are two cases. Either M is a (τ,a)-structure, or M is not. In the first case there are ordinals ν_1, ν_2 such that $\nu_1 \leq \nu_2$, and for all $i \in I$, all finite lists b from I_M: $|\{i\}(b)| < \zeta \iff |\{i\}_M(b)|_M < \nu_1$, $|\{i\}(b)| < \tau \iff |\{i\}_M(b)|_M < \nu_2$. There are two subcases: $\nu_1 = \nu_2$, $\nu_1 < \nu_2$.

In the second subcase there is an index i and a list b from I_M such that $|\{i\}_M(b)|_M = \nu_1$. P is definable inside M with parameters a, b by: $n \in P \iff |\{n\}_M(a)|_M < |\{i\}_M(b)|_M$. Hence $P \in I_M$. $|\{e_0\}_M(a,P)|_M < \nu_2$ because $|\{e_0\}_M(a,P)|_M < \tau$. Also $\exists y \in I_M |\{e\}_M(y,a)|_M < |\{e_0\}_M(a,P)|_M$ because $\exists y |\{e\}(y,a)| < |\{e_0\}(a,P)|$ and M is a (τ,a)-structure.

In the first subcase $\sup\{|\{i\}_M(b)|_M : \{i\}_M(b)\downarrow\} = \nu_1$. Obviously the supremum $\geq \nu_1$ because the set contains all $|\{i\}_M(a)|_M$ with $|\{i\}(a)| < \zeta$. If the supremum was strictly greater than ν_1 then

$P \in I_M$ by the same proof as above. Choose an i such that $|\{i\}(a,P)| = \zeta$. Then $|\{i\}(a,P)| < \tau$, $|\{i\}_M(a,P)|_M < \nu_2$. $|\{i\}_M(a,P)|_M \geq \nu_1$ because $|\{i\}(a,P)| \geq \zeta$. Hence $\nu_1 < \nu_2$, a contradiction.

The proof goes as follows. Find indexes e_1, e_2, e_3 such that the statements in the following table are true, when x is a code for M:

Case II : $|\{e_1\}(x,a)| < \tau$.

Case I : $|\{e_1\}(x,a)| \geq \tau$.

" subcase I : $|\{e_2\}(x,a)| = \zeta$.

" " II : $|\{e_2\}(x,a)| > \zeta$, $|\{e_3\}(x,a)| < \tau$, and for some y $\{e\}(y,a)$ is a subcomputation of $\{e_3\}(x,a)$.

Let $\{e_4\}(a)$ be the following computation:
$\forall x$ (x is the code for a countable recursion structure $\Rightarrow \{e_1\}(x,a)\downarrow$ or $\{e_2\}(x,a)\downarrow$ or $\{e_3\}(x,a)\downarrow$).
By lemma 39 and the properties in the table $|\{e_4\}(a)| \leq \tau$. Hence $|\{e_4\}(a)| < \zeta$ because $\tau < \xi$ and $|\{e_4\}(a)|$ is a-constructive. By lemma 42 there is an x which brings us to case I, subcase II. For this x $|\{e_1\}(x,a)| > \zeta$ and $|\{e_2\}(x,a)| > \zeta$. The length of the computation "$\{e_1\}(x,a)\downarrow$ or $\{e_2\}(x,a)\downarrow$ or $\{e_3\}(x,a)\downarrow$" is $\inf\{|\{e_1\}(x,a)|, |\{e_2\}(x,a)|, |\{e_3\}(x,a)|\}$ + a finite number. As $|\{e_4\}(a)| < \zeta$ this infimum is less that ζ. Hence $|\{e_3\}(x,a)| < \zeta$. By the table there is a y such that $\{e\}(y,a)$ is a subcomputation of $\{e_3\}(x,a)$. Hence $\exists y |\{e\}(y,a)| < \zeta$.

Let e_1 be the index from lemma 43.

Construction of e_2. Let $\{e_2\}(x,a)$ be the following computation: For all $i \in N$, for all finite lists b from I_M, if $\{i\}_M(b)\downarrow$ then compute $\{i\}(b)$.

Construction of e_3. Let $\{e_3\}(x,a)$ be the following computation: Search through I_M to find a z such that $\{e_o\}(a,z)\downarrow$, and $\exists y|\{e\}(y,a)| < |\{e_o\}(a,z)|$. Since M is coded by x the search through I_M can be expressed as a search through N. In case I, subcase II, there is a z in I_M which satisfies this claim, namely $z = P$. By several lemmas in § 10 we see that $|\{e_3\}(x,a)| = |\{e_0\}(a,z)| +$ a finite number, where z is chosen such that $\exists y |\{e\}(y,a)| < |\{e_0\}(a,z)|$, and $|\{e_0\}(a,z)|$ is minimal. $z = P$ gives $|\{e_0\}(a,P)| < \tau$. Hence $|\{e_3\}(x,a)| < \tau$. Also $\{e\}(y,a)$ will be a subcomputation of $\{e_3\}(x,a)$ for some y because of the claim $\exists y |\{e\}(y,a)| < |\{e_0\}(a,z)|$. □

Corollary 1: Suppose $[\zeta,\xi)$ is an a-gap such that $\zeta + \omega < \xi$. Then $\zeta + \omega$ is a-reflecting.

Proof: By theorem 14 it is enough to prove that $\zeta + \omega \leq \zeta^P$. Let $\{e\}(a,z)$ be the following computation: For all n in the set with characteristic function z compute $\{n\}(a)$. Then $\{e\}(a,P)\downarrow$, and $|\{e\}(a,P)| = \sup\{|\{n\}(a)| : n \in P\} = \zeta$. Hence $\zeta < \zeta^P$. As ζ^P is a limit ordinal $\zeta + \omega \leq \zeta^P$. □

Let $\xi^- = \xi$ if ξ is a limit of limit ordinals, $= \tau$ if $\xi = \tau + \omega$ where τ is a limit. Let $\xi' = (\xi^-)^-$.

Corollary 2: Suppose $[\zeta,\xi)$ is an a-gap. Then $\inf\{\zeta + \aleph_1, \xi'\}$ is a-reflecting.

Proof: To avoid the trivial case suppose $\zeta + \omega + \omega < \xi$. Suppose $\exists y |\{e\}(y,a)| < \inf\{\zeta + \aleph_1, \xi'\}$. By theorem 14 it is enough to prove that $\exists y |\{e\}(y,a)| < \zeta^P$, i.e. to find an index e_0 such that $\exists y |\{e\}(y,a)| < |\{e_0\}(a,P)|$, and $|\{e_0\}(a,P)| \leq \inf\{\zeta + \aleph_1, \xi'\}$. The computation $\{e_0\}(a,P)$ will involve the set WO of prewellorderings. For each $w \in WO$ we decide whether or not $\exists y |\{e\}(y,a)| < \zeta + |w|$ by a computation with length $\leq \max(\zeta,\eta) +$ a finite number, where $\eta = \min(\zeta + |w|, \inf\{|\{e\}(y,a)| : y \in I\})$. As $\exists y |\{e\}(y,a)| < \zeta + \aleph_1$ it will be the case that $\exists y |\{e\}(y,a)| < \zeta + |w|$ for some $w \in WO$. Hence $|\{e_0\}(a,P)| \leq \max(\zeta, \inf\{|\{e\}(y,a)| : y \in I\}) + \omega$, hence $|\{e_0\}(a,P)| \leq$

$\inf\{\zeta + \aleph_1, \xi'\}$. It will also be the case that $\exists y |\{e\}(y,a)| < |\{e_o\}(a,P)|$.

A more detailed construction of e_o : Choose a computation $\{e_1\}(a,P)$ such that $|\{e_1\}(a,P)| = \zeta$ (let for instance $\{e_1\}(a,y) \simeq 0$ if $\{n\}(a)\downarrow$ for all n in the set with characteristic function y). By applications of the recursion theorems one can find an index e_2 such that $\{e_2\}(w,i,t,a,P)\downarrow$ iff $w \in WO$ and i is in the domain of w, in which case

$$\{e_2\}(w,i,t,a,P) \simeq \begin{cases} 0 & \text{if } \zeta + |i|_w < |t| \\ 1 & \text{if } \zeta + |i|_w \geq |t| , \end{cases}$$

and $|\{e_2\}(w,i,t,a,P)| = \max(\zeta, \min\{\zeta + |i|_w, |t|\})$ + a finite number which depends only on the number of inputs in the computation t, and on k, where $\min\{\zeta + |i|_w, |t|\}$ = a limit ordinal + k, $k < \omega$. To express the ordinal ζ we use the computation $\{e_1\}(a,P)$. Hence P appears in the input list of $\{e_2\}$, and $\zeta < |\{e_2\}(w,i,t,a,P)|$. Choose e_3 such that

$$\{e_3\}(w,i,a,P) \simeq \begin{cases} 0 & \text{if } \forall y \{e_2\}(w,i,\langle e,y,a\rangle,a,P) \simeq 0 \\ 1 & \text{if } \exists y \{e_2\}(w,i,\langle e,y,a\rangle,a,P) \simeq 1 . \end{cases}$$

The quantifiers are expressed by E. If $\forall y(\zeta + |i|_w < |\{e\}(y,a)|)$ then $\{e_3\}(w,i,a,P) \simeq 0$. If $\exists y(\zeta + |i|_w \geq |\{e\}(y,a)|)$ then $\{e_3\}(w,i,a,P) \simeq 1$. In any case $|\{e_3\}(w,i,a,P)| = \zeta + |i|_w$ + a finite number which depends only on k', where $\zeta + |i|_w$ = a limit ordinal + k', $k' < \omega$. Hence $\exists y |\{e\}(y,a)| < |\{e_3\}(w,i,a,P)|$ if $\{e_3\}(w,i,a,P) \simeq 1$. Choose e_4 such that

$$\{e_4\}(w,a,P) \simeq \begin{cases} 0 & \text{if } \forall i \in \operatorname{dom} w \ \{e_3\}(w,i,a,P) \simeq 0 \\ 1 & \text{if } \exists i \in \operatorname{dom} w \ \{e_3\}(w,i,a,P) \simeq 1 \end{cases}$$

The universal quantifier is expressed by E, and the existential quantifier by a selection operator for numbers. If $\forall y(\zeta + |w| \leq |\{e\}(y,a)|)$ then $\{e_4\}(w,a,P) \simeq 0$, and $|\{e_4\}(w,a,P)| = \zeta + |w|$ + a finite number which depends only on k'', where $|w|$ = a limit ordinal + k'', $k'' < \omega$. If $\exists y(\zeta + |w| > |\{e\}(y,a)|)$ then $\{e_4\}(w,a,P) \simeq 1$, and

$|\{e_4\}(w,s,P)| = \max\{\zeta, |\{e_3\}(w,i,a,P)|\}$ + a finite number, where i is chosen such that $\{e_3\}(w,i,a,P) \simeq 1$ and $|\{e_3\}(w,i,a,P)|$ is minimal. By choosing i such that $\zeta + |i|_w = \inf\{|\{e\}(y,a)| : y \in I\}$ we see that $|\{e_4\}(w,a,P)| \leq \max\{\zeta, \inf\{|\{e\}(y,a)| : y \in I\}\}$ + a finite number, and also $\exists y |\{e\}(y,a)| < |\{e_4\}(w,a,P)|$. So in any case $|\{e_4\}(w,a,P)| < \inf\{\zeta + \aleph_1, \xi'\}$. Let $\{e_0\}(a,P)$ be the computation: For all $w \in WO$ compute $\{e_4\}(w,a,P)$. Then $|\{e_0\}(a,P)| \leq \inf\{\zeta + \aleph_1, \xi'\}$. For all $w \in WO$ $|\{e_4\}(w,a,P)| < |\{e_0\}(a,P)|$. Choosing a $w \in WO$ such that $\exists y |\{e\}(y,a)| < \zeta + |w|$ we have that $\exists y |\{e\}(y,a)| < |\{e_4\}(w,a,P)|$, hence $\exists y |\{e\}(y,a)| < |\{e_0\}(a,P)|$. □

<u>Remark</u>: Suppose η is an a-constructive ordinal, and $[\zeta, \xi)$ is an a-gap such that $\zeta < \eta + \aleph_1$. Then $\inf\{\zeta + \aleph_1, \xi^-\}$ is a-reflecting. (When the length of the gap is countable and $\xi = \lambda + \omega + \omega, \lim \lambda$, this conclusion is slightly stronger than the conclusion of cor. 2.)

Proof: Suppose $\exists y |\{e\}(y,a)| < \inf\{\zeta + \aleph_1, \xi^-\}$. Let τ be a limit ordinal such that $\exists y |\{e\}(y,a)| < \tau$, $\tau < \xi$. Define e_2 as in the proof of corollary 2, starting from level η instead of ζ. Since η is a-constructive P is not in the input list of $\{e_2\}$. Define e_3, e_4 and e_0 from e_2 as in the proof of corollary 2. Then $\exists y |\{e\}(y,a)| < |\{e_0\}(a)|$, and $|\{e_0\}(a)| \leq \tau$. Hence $|\{e_0\}(a)| < \zeta$ since there are no a-constructive ordinals in the a-gap. Hence $\exists y |\{e\}(y,a)| < \zeta$. □

<u>Corollary 3</u>: Suppose $[\zeta, \xi)$ is an a-gap, and suppose that there is a prewellordering with length α which is recursive in a below ζ, such that $\zeta + \alpha + \omega < \xi$. Then $\zeta + \alpha$ is a-reflecting.

Proof: Suppose $\exists y |\{e\}(y,a)| < \zeta + \alpha$. By theorem 14 it is enough to find an index e_0 such that $\exists y |\{e\}(y,a)| < |\{e_0\}(a,P)|$, and $|\{e_0\}(a,P)| < \zeta + \alpha + \omega$. e_0 is constructed in almost the same way as in corollary 2.

First choose e_1 such that $|\{e_1\}(a,P)| = \zeta$. Let the prewellordering with length α be denoted by R. If $u \in \mathrm{dom}(R)$, let $|u|_R$ be the ordinal of u in R. If $u \notin \mathrm{dom}(R)$ let $|u|_R = \alpha$. There is an index e_2 such that

$$\{e_2\}(u,t,a,P) \simeq \begin{cases} 0 & \text{if } \zeta+|u|_R < |t| \\ 1 & \text{if } \zeta+|u|_R \geq |t|, \end{cases}$$

and $\inf\{\zeta+|u|_R, |t|\} < |\{e_2\}(u,t,a,P)| < \inf\{\zeta+|u|_R, |t|\}$ + a finite number which depends only on the number of inputs in the computation t, and on k, where $\inf\{\zeta+|u|_R, |t|\}$ = a limit ordinal + k, $k < \omega$. There is an index e_3 such that

$$\{e_3\}(u,a,P) \simeq \begin{cases} 0 & \text{if } \forall y\{e_2\}(u,\langle e,y,a\rangle,a,P) \simeq 0 \\ 1 & \text{if } \exists y\{e_2\}(u,\langle e,y,a\rangle,a,P) \simeq 1. \end{cases}$$

Hence

$$\{e_3\}(u,a,P) \simeq \begin{cases} 0 & \text{if } \forall y(\zeta+|u|_R < |\{e\}(y,a)|) \\ 1 & \text{if } \exists y(\zeta+|u|_R \geq |\{e\}(y,a)|). \end{cases}$$

$|\{e_3\}(u,a,P)| = \zeta + |u|_R$ + a finite number which depends on k', where $\zeta + |u|_R$ = a limit ordinal + k', $k' < \omega$. By choosing $|u|_R$ large enough we have $\exists y |\{e\}(y,a)| < |\{e_3\}(u,a,P)|$. Let $\{e_0\}(a,P)$ be the following computation: For all $u \in \mathrm{dom}(R)$ compute $\{e_3\}(u,a,P)$. Then $|\{e_0\}(a,P)| = \zeta + \alpha$ + a finite number, and $\exists y |\{e\}(y,a)| < |\{e_0\}(a,P)|$. □

<u>Definition</u>: A set A is recursively enumerable in a below ζ if there is an index i such that for all x: $x \in A \Longleftrightarrow |\{i\}(x,a)| < \zeta$.

<u>Theorem 15</u>: Suppose $[\zeta,\xi)$ is an a-gap such that $\zeta + \omega$ is a-reflecting (by cor. 1 this is true if the length of the gap is larger than ω). Suppose that the set A is recursively enumerable in a below ζ, and that A contains the characteristic function of a subset of the natural numbers which is recursively enumerable in a below ζ. Then A contains the characteristic function of a subset of the natural numbers

which is recursive in a below ζ.

Proof: Let e_0 be an index such that for all y: $y \in A \iff |\{e_0\}(y,a)| < \zeta$. Suppose $u \subseteq N$, $u \in A$, u is recursively enumerable in a below ζ. Let e_1 be an index such that for all natural numbers m: $m \in u \iff |\{e_1\}(m,a)| < \zeta$. If x is a convergent computation let $u^x = \{m : |\{e_1\}(m,a)| < |x|\}$. There is an index e_2 such that $\{e_2\}(m,a,x)\downarrow$ iff x is convergent, in which case $\lambda m\{e_2\}(m,a,x)$ is the characteristic function of u^x, and $|x| < |\{e_2\}(m,a,x)| < |x|$ + a finite number which is independent of m.

Let $\{e\}(x,a)$ be the following computation:
Substitute $\lambda m\{e_2\}(m,a,x)$ for y in $\{e_0\}(y,a)$, and compute $\{e_0\}(y,a)$. Then $\{e\}(x,a)\downarrow$ iff x is convergent and $u^x \in A$. If $\{e\}(x,a)\downarrow$ then $|\{e\}(x,a)| = \sup(\{|\{e_0\}(u^x,a)|\} \cup \{|\{e_2\}(m,a,x)| : m \in N\})$ + a finite number.

Choose an index e_3 such that $|\{e_3\}(a,P)| = \zeta$. Let $x = \langle e_3,a,P\rangle$. Then $u^x = u$, $\{e\}(x,a)\downarrow$, $|\{e_0\}(u^x,a)| < \zeta$ because $u^x \in A$, $|\{e_2\}(m,a,x)| < \zeta$ + a finite number independent of m. Hence $|\{e\}(x,a)| < \zeta + \omega$. By assumption $\zeta + \omega$ is a-reflecting. Hence $\exists x |\{e\}(x,a)| < \zeta$. Choose e_4 such that $|\{e_4\}(a)| < \zeta$ and $\exists x |\{e\}(x,a)| < |\{e_4\}(a)|$.

Let $B = \{x : |\{e\}(x,a)| < |\{e_4\}(a)|\}$, and let $C = \{x : x \in B$ and $|x|$ is minimal$\}$. Then B and C are nonempty, and both are recursive in a below ζ. If $x \in B$ then $|x| < \zeta$ because $|x| < |\{e\}(x,a)|$. Let $x \in C$. Then $\{e\}(x,a)\downarrow$, hence $u^x \in A$. It remains to prove that u^x is recursive in a below ζ.

If $y,z \in C$ then $|y| = |z|$, hence $u^y = u^z$. $\lambda m E(\lambda y\{e_5\}(m,a,y))$ is the characteristic function of u^x, where $\{e_5\}(m,a,y)$ is the following computation: First see if $y \in C$. If not let $\{e_5\}(m,a,y) \simeq 1$. If $y \in C$ (then $u^y = u^x$) let $\{e_5\}(m,a,y) \simeq \{e_2\}(m,a,y)$. When $y \in C$ $|\{e_2\}(m,a,y)| < |y| + \omega < \zeta$. Hence $|\{e_5\}(m,a,y)| < \zeta$ for all m, y. Hence u^x is recursive in a below ζ. □

In theorem 16 we prove that in some a-gaps there are ordinals which are not a-reflecting.

<u>Theorem 16</u>: Suppose $[\zeta,\xi)$ is an a-gap such that $\zeta = \gamma(\alpha,\beta)$, $\alpha = \tau+\omega$, $\lim \tau$, τ is recursive in a below ζ, and there is a $w \in WO$ such that $\beta = |w|$ and w is recursive in a below ζ. Then there are ordinals in the segment $(\zeta+\tau,\xi)$ which are not a-reflecting.

Proof: α is recursive in a below ζ because τ is. By theorem 14 the length of the gap is α, i.e. $\zeta+\tau+\omega = \xi$. We will find an index e such that $\exists y|\{e\}(y,a)| < \xi$, and $\forall y|\{e\}(y,a)| > \zeta+\tau$. If y is a computation such that $|y| = \zeta$, then the following statement is true: "The order type of the a-gaps below $|y|$ with length $\geq \alpha$ is β, and $|y|$ is in an a-gap with length $\geq \tau$". Let $\{e\}(y,a)$ be the computation which expresses this fact. If $\{e\}(y,a)\downarrow$ then $\zeta \leq |y| < \xi$. By results in § 10 $\zeta+\tau < |\{e\}(y,a) < \xi$ when $\{e\}(y,a)\downarrow$. Hence $\exists y|\{e\}(y,a)| < \xi$, and $\forall y|\{e\}(y,a)| > \zeta+\tau$. □

<u>Corollary</u>: Let $[\zeta,\xi)$ be the first a-gap with length $\tau+\omega$, where τ is one of the following ordinals: 0, $\omega\cdot n$, ω^n, $\aleph_1$. Then $\zeta+\tau$ is a-reflecting, and ξ $(=\zeta+\tau+\omega)$ is not.

Proof: $\zeta+\tau$ is trivially a-reflecting when $\tau = 0$, it is a-reflecting by the remark following corollary 2 of theorem 14 when $\tau = \omega\cdot n$, it is a-reflecting by corollary 2 of theorem 14 when $\tau = \omega^n$, $\tau = \aleph_1$. ξ is not a-reflecting by theorem 16. □

Let ζ and τ be ordinals. <u>τ is a-reflecting to ζ</u> if for all natural numbers e: $\exists x|\{e\}(x,a)| < \tau \Rightarrow \exists x|\{e\}(x,a)| < \zeta$. (Hence η is a-reflecting iff η is a-reflecting to $\sup\{\nu+1 : \nu$ is a-constructive, $\nu < \eta\}$.)

If ζ is an ordinal let ζ_r^a be the greatest ordinal μ such that μ is a-reflecting to ζ. If ζ is a-constructive then $\zeta_r^a = \zeta$.

If $[\zeta,\xi)$ is an a-gap then $\zeta_r^a \leq \xi$ because $\xi+1$ is not a-reflecting to ζ. If $[\zeta,\xi)$ is the first a-gap then $\zeta_r^a < \xi$ by the corollary of theorem 16. If $[\zeta,\xi)$ is the first a-gap with length ω^2 then ξ is a-reflecting by theorem 14. Hence $\zeta_r^a = \xi$.

Theorem 17: Suppose $[\zeta,\xi)$ is an a-gap, and suppose A is nonempty and corecursively enumerable in a below ζ. Suppose that there is a limit ordinal γ such that $\zeta_r^a \leq \gamma < \xi$. Then there is an $x \in A$ such that the segment $[\zeta_r^a,\gamma)$ is contained in an a,x-gap.

Proof: Since ζ_r^a is the greatest ordinal which is a-reflecting to ζ there is an index e_o such that $\exists y|\{e_o\}(y,a)| = \zeta_r^a$, $\forall y|\{e_o(y,a)| \geq \zeta_r^a$. Let e_1 be an index such that $x \notin A \iff |\{e_1\}(x,a)| < \zeta$. To obtain a contradiction suppose that for all x in A there is an index e such that $\zeta_r^a \leq |\{e\}(x,a)| < \gamma$. Then the following is true: $\forall x[x \notin A$ or $\exists e(\{e\}(x,a)\downarrow$ and $\zeta_r^a \leq |\{e\}(x,a)|)]$. Let $\{e_2\}(a)$ be the following computation: $\forall x[\{e_1\}(x,a)\downarrow$ or $\exists e(\{e\}(x,a)\downarrow$ and $\exists y(|\{e_o\}(y,a)| \leq |\{e\}(x,a)|))]$. Let x be arbitrary. The length of the computation inside the brackets [ ] is $\inf\{|\{e_1\}(x,a)|, \inf\{|\{e\}(x,a)|: \exists y(|\{e_o\}(y,a)| \leq |\{e\}(x,a)|)\}\}$ + a finite number. Hence the length is less than γ. Hence $|\{e_2\}(a)| \leq \gamma$. Since there is no a-constructive ordinal in the gap $|\{e_2\}(a)| < \zeta$. Let $x \in A$. Then $|\{e_1\}(x,a)| \geq \zeta$. Hence $\inf\{|\{e\}(x,a)|: \exists y(|\{e_o\}(y,a)| \leq |\{e\}(x,a)|)\} < \zeta$. Hence $\exists y|\{e_o\}(y,a)| < \zeta$, a contradiction. □

Remark: Suppose $[\zeta_r^a,\gamma)$ is contained in an a,x-gap $[\mu,\nu)$. Then $\mu_r^{a,x} \leq \zeta_r^a$, because if $\mu_r^{a,x} > \zeta_r^a$ then by reflection $\exists y|\{e_o\}(y,a)| < \mu$, a contradiction.

Remark: Theorems 14, 15 and 17 are analogues to theorem 2.2, corollary 2.3 and lemma 2.5 respectively in Harrington [7].

Part C. To characterize $\gamma(\omega,0)$.

In this section we want to discuss some of the properties of the first ordinal in the first gap. We let the sets of subindividuals be ω, the natural numbers, and the set I of individuals be ${}^{\omega}\omega$, Baire-space. We will then deal with recursion in the type-3 functional ${}^{3}E$. Recall that $\gamma(\alpha,\beta)$ is the first ordinal in the β th gap of length $\geq \alpha$ starting the enumeration with 0. Thus we are here interested in $\gamma(\omega,0)$.

We will give two characterizations of $\gamma(\omega,0)$, one by the reflection properties of the ordinal, the other in a definability-theoretic way. Other results are just giving properties of $\gamma(\omega,0)$ and several other first ordinals in gap's.

Some of the proofs use additional axioms to ZFC. Other proofs are just briefly sketchedout. In those cases a formal proof will need machinery not available to us. The proofs are, however, not difficult given the machinery. In those cases we will refer to suitable sources for background material.

<u>Definition</u>:

<u>a</u> We say that an ordinal γ is <u>strictly countable</u> (in ${}^{3}E$) if for any $\beta < \gamma$ there is a $w \in WO$ such that $\|w\| = \beta$ and w is recursive in ${}^{3}E$ below γ.

<u>b</u> A normal recursion theory Θ satisfies <u>WO-uniformization</u> if for any limit ordinal τ and set $A \subseteq WO$, if A is an initial segment of WO recursive below τ, then there is a $w \in WO$ recursive below τ such that $\|w\| \geq \|A\|$.

<u>Remarks</u>: 1. Any strictly countable ordinal will be countable. Moreover, if recursion in ${}^{3}E$ satisfies WO-uniformization, all countable ordinals that are beginnings of gaps will be strictly countable. We leave the proof of this fact as an exercise for the reader.

2. The only way known to the authors of obtaining WO-uniformization

is via ordinary uniformization, i.e. an effective way of picking an element out of a recursive, nonempty set. However, we still regard the possibility that WO-uniformization may always hold as open. That it sometimes holds follows from the following.

Observations:

a If there is, for some n, a Π^1_n wellordering of ${}^\omega\omega$, ordinary uniformization holds (ex. $V = L$, $V = L[0^\#]$, $V = L^\mu$).

b If all sets recursive in ${}^3E$ and some individual are determined, then ordinary uniformization holds.

The proof of a is a simple exercise while in order to prove b we need some background from e.g. Moschovakis [25] or other recent presentations of modern descriptive set theory. Using techniques of Moschovakis and the recursion theorem we prove that any set recursive in ${}^3E$ and some individual has a recursive in ${}^3E$ scale (not defined here), and thus satisfies uniformization.

WARNING: This sketch does not work for any other functional than ${}^3E$.

We will now start the definability-theoretic characterization.

Definition: Let $\mathfrak{A} = \langle A_1, A_2, \mathcal{P}, \in \rangle$ where A_1 is a transitive set, $A_1 \subseteq A_2$, $\mathcal{P}: A_1 \to A_2$ is the power set operator and $\in$ the standard 'element in' relation.

We say that $\mathfrak{A}$ is 2nd order admissible if

1. $\omega \in A_1$ and $\forall y(x \in A_1 \Rightarrow x \cap y \in A_1)$
2. A_i is rudimentary closed ($i = 1.2$) (see e.g. Devlin [27] for definition).
3. Δ_0-collection: Let φ be a Δ_0-formula with parameters $\vec{x}$ from A_i. Suppose

   $$\forall y \in x\ \exists z \in A_i \varphi(y, z, \vec{x})$$

   Then there is a $u \in A_i$ such that

   $$\forall y \in x\ \exists z \in u\, \varphi(y, z, \vec{x})$$

We see that A_1 will be admissible and that we will have Σ_1-collection and Δ_1-separation over $\mathfrak{A}$.

We will now construct a hierarchy for 2nd order admissible structures; let HF mean Heriditarily Finite. $\mathfrak{A}^0 = \langle HF, HF, \mathcal{P}, \in\rangle$

If λ is a limit ordinal, let $A_i^\lambda = \bigcup_{\delta<\lambda} A_i^\delta$; $\underline{i = 1,2}$. The construction of $\mathfrak{A}^{\alpha+1}$ from $\mathfrak{A}^\alpha$ is by an inductive definition over ω:

$$A_1^{\alpha,0} = A_1^\alpha \cup \{A_1^\alpha\}, \quad A_2^{\alpha,0} = A_2^\alpha \cup \{A_2^\alpha\} \cup \mathcal{P}(A_1^\alpha)$$

$$A_1^{\alpha,k+1} = A_1^{\alpha,k} \cup \text{Rudimentary closure of } A_1^{\alpha,k} \cup \{z \cap y;\ z \in A_1^{\alpha,k} \,\&\, y \in A_2^{\alpha,k}\}.$$

$$A_2^{\alpha,k+1} = A_2^{\alpha,k} \cup \text{Rudimentary closure of } A_2^{\alpha,k} \cup \{\mathcal{P}(z);\ z \in A_1^{\alpha,k+1}\}$$

$$A_1^{\alpha+1} = \bigcup_{k\in\omega} A_1^{\alpha,k} \quad \text{and} \quad A_2^{\alpha+1} = \bigcup_{k\in\omega} A_2^{\alpha,k}$$

$A_1^{\alpha+1}$ and $A_2^{\alpha+1}$ are the mimimal extensions of $A_1^\alpha \cup \{A_1^\alpha\}$, $A_2^\alpha \cup \{A_2^\alpha\}$ that satisfy 1. and 2. in the definition of 2nd order admissibility.

<u>Definition</u>: An ordinal γ is 2nd order admissible if $\mathfrak{A}^\gamma$ is 2nd order admissible.

<u>Remark</u>: To see that the first 2nd order admissible ordinal is countable, we may use a standard method of producing a notation system via a Gödel-enumeration of the formulas.

<u>Lemma 45</u>: (Inductive definability) Let $\mathfrak{A}$ be 2nd order admissible. Let $\Delta : A_2 \to A_2$ be a function whose graph is Δ_1-definable over $\mathfrak{A}$. Let $x \in A_2$. Define $\Gamma : A_1 \to A_2$ by

$$\Gamma(\emptyset) = x$$

$$\Gamma(y) = \Delta(\langle\Gamma(z)\rangle_{z\in y}) \quad (= \Delta(\Gamma \upharpoonright y))$$

Then Γ is Δ_1-definable over $\mathfrak{A}$, and for $y \in A_1$ we have $\Gamma(y) \in A_2$.

The proof is standard and we give a brief outline: By induction on the rank of y we prove

<u>i</u> $x = \Gamma(y) \iff \exists f \in A_2$ (f is a function defined on the transitive closure of $y \cup \{y\}$ and $\forall z (z = \emptyset \Rightarrow f(z) = x$ & $(z \neq \emptyset$ & $f(z)$ is defined $\Rightarrow f(z) = \Delta(f | z))$ & $x = f(y)$)

$\iff \forall f \in A_2$ (f is a function etc. $\Rightarrow x = f(y)$).

<u>ii</u> $\Gamma(y)$ and $\Gamma | y$ are both elements of A_2. □

<u>Lemma 46</u>: Let $\mathcal{B} = \langle B_1, B_2, \mathcal{P}, \in \rangle$ be a 2nd order admissible structure. The function $\Gamma(\alpha) = \mathfrak{A}^\alpha$ is Δ_1-definable over $\mathcal{B}$, and if $\alpha \in B_1$ then $\mathfrak{A}^\alpha \in B_2$.

Proof: The construction of $\mathfrak{A}^{\alpha+1}$ from $\mathfrak{A}^\alpha$ is Δ_1. That $\mathfrak{A}^\alpha \in B_2 \Rightarrow \mathfrak{A}^{\alpha+1} \in B_2$ is trivial by collection, so by lemma 45 the claim follows. □

<u>Corollary</u>: Let γ be the least 2nd order admissible ordinal. Then $\mathfrak{A}^\gamma$ is the least 2nd order admissible structure.

<u>Theorem 18</u>:

<u>a</u> Let γ be strictly countable and the beginning of a gap. Then γ is 2nd order admissible.

<u>b</u> Let γ be 2nd order admissible. Then there is no computation in ${}^3E$ with arguments from ω of length γ.

<u>Corollary</u>: A strictly countable ordinal is the beginning of a gap if and only if it is 2nd order admissible.

We must content ourselves with a sketch of the proof. Background material is found in Normann [19] and Sacks [21].

<u>Sketch of proof of a</u>: Let $w \in WO$. Uniformly in w we may find <u>codes</u> (not defined here) for $A_1^{\|w\|}$, $A_2^{\|w\|}$ resp., uniformly recursive in w, ${}^3E$, where the code for $A_1^{\|w\|}$ is a code in the sense of Sacks [20], the code $A_2^{\|w\|}$ is a code in the sense of Normann [19] or Sacks [21]. The

computation of these codes should not be 'too long' compared with $\|w\|$. What we want to prove is: If φ is Σ_1 and $x \in \mathfrak{A}^\gamma$ and

(1) $\forall y \in x \ \ \exists\alpha < \gamma \ \ \mathfrak{A}^\alpha \models \varphi(y)$, then

(2) $\exists\alpha < \gamma \ \ \forall y \in x \ \ \mathfrak{A}^\alpha \models \varphi(y)$.

By the coding method and Gandy's selection, (1) may be translated to the totality below γ of a certain recursive function defined on a code for x. Since γ begins a gap, the whole function must be recursive below γ, which translates back to (2).

Proof of b: By lemma 45 the following function is Δ_1-definable over $\mathfrak{A}^\gamma$:

$\Gamma(\alpha)$ = The tree of computations in ${}^3E$ of length $\leq \alpha$.

Now, if ${}^3E(\lambda x\{e\}(x,\vec{n}))$ has length $\leq \gamma$, then

$\forall x \in I \ \ \exists\alpha < \gamma(\{e\}(x,\vec{n})$ is a computation of length $\leq \gamma)$.

By collection in $\mathfrak{A}^\gamma$ we see that the computation itself is shorter than γ. This proves b. □

S.S. Wainer has studied a concept similar to 2nd order admissibility. The corollary to theorem 18 gives a weak 3rd order characterization of $\gamma(\omega,0)$. We will now prove that it is impossible to obtain a 2nd order characterization.

Theorem 19 (2nd order reflection): Let γ be a 2nd order admissible ordinal that is not a cardinal number. Let φ be any second order formula in the language of set theory, with parameters from L_γ. If $L_\gamma \models \varphi$, there will be a $\beta < \gamma$ such that $L_\beta \models \varphi$.

Remark: We will prove the theorem for countable γ. The general proof is not more complicated. By the same idea, we may prove directly that the countable beginning of a gap will reflect 2nd order formulas.

Proof: Assume that $L_\gamma \models \varphi$.

$\{A ; A \subseteq \omega \times \omega \,\&\, A$ is isomorphic to an initial segment of $L\}$ is 1st order definable over $\mathcal{P}(\omega)$.

Let $B = \{A \subseteq \omega \times \omega ;\ A$ is isomorphic to the least intial segment L_α of L such that $L_\alpha \models \varphi\}$

Then $B \in A_2^\omega$, so $B \in A_2^\gamma$.

Since $L_\gamma \models \varphi$ we have

$$\forall a \in \omega\ \forall A \in B\, \exists \beta < \gamma \ \text{(If } a \text{ codes an ordinal } \|a\|_A \text{ in } A, \text{ then } \beta = \|a\|_A)$$

The codes for ordinals in A will be $\Delta_o(A)$-definable, and the relation $\beta = \|a\|_A$ is $\Delta_o(\beta, \mathcal{P}(\beta \times \omega), A, a)$-definable.

By collection in $\mathfrak{A}^\gamma$, we find an $\alpha_o < \gamma$ such that

$$\forall a \in \omega\ \forall A \in B\, \exists \beta < \alpha_o \ \text{(If } a \text{ codes an ordinal } \|a\|_A \text{ in } A, \text{ then } \beta \models \|a\|_A)$$

It follows that the least β such that $L_\beta \models \varphi$ is less than γ. □

The reflection properties reported on in Theorem 19 is not sufficient to characterize the beginning of a gap.

<u>Definition</u>: Let γ be any admissible ordinal. By some standard coding of formulas, let $\mathcal{L}_\gamma^i$ be the language of i'th order of set theory that is closed under arbitrary conjunctions and disjunctions over elements of L_γ. We will identify a formula of $\mathcal{L}_\gamma^i$ with its code in L_γ.

<u>Lemma 47</u>: Let γ be a countable 2nd order admissible ordinal. Let $\varphi(X_1,\dots,X_n,x_1,\dots,x_k)$ be any formula in $\mathcal{L}_\gamma^2$ with second order variables X_i and first order variables x_j. Then the following relation is Δ_1-definable over $\mathfrak{A}^\gamma$:

$\{A, A_1,\dots,A_n \subseteq \omega,\ a_1,\dots,a_k \in \omega ;\ A$ codes an initial segment L_β of L and $A_i \subseteq A$ codes a subset X_i of L_β and a_j codes an element x_j of L_β $(i = 1,\dots,n ;\ j = 1,\dots,k)$ and $L_\beta \models \varphi(X_1,\dots,X_n,x_1,\dots,x_k)\}$.

Proof: By induction on rank φ, this is an immediate consequence of Lemma 45. □

Theorem 20:

a Let γ be a countable 2nd order admissible ordinal. Let φ be any closed formula in $\mathcal{L}^2_\gamma$, and assume $L_\gamma \models \varphi$. Then for some $\beta < \gamma$, $L_\beta \models \varphi$.

b Let γ be as in a. Let $\Gamma : \gamma \to \mathcal{L}^2_\gamma$ be a Δ_1-function over L_α mapping α on a formula φ_α with one free variable. Assume that for some $x \in L_\gamma$:

$$\forall y \subseteq x \, \exists \alpha \, L_\gamma \models \varphi_\alpha(y)$$

Then $\exists \beta < \gamma \, \forall y \subseteq x \, \exists \alpha_1 < \beta \, \exists \alpha_2 < \beta \, L_{\alpha_1} \models \varphi_{\alpha_2}(y)$.

c Let γ be as in a, and let $B - \mathcal{L}^2_\gamma$ be the fragment of $\mathcal{L}^2_\gamma$ where all quantifiers are bounded. ($\forall y \subseteq x$ and $\forall y \in x$). Let $\Gamma : \gamma \to B - \mathcal{L}^2_\gamma$ and x be as in b. Then $\exists \beta < \gamma \, \forall y \subseteq x \, \exists \alpha < \beta \, L_\beta \models \varphi_\alpha(y)$.

d Let γ be admissible. If for any Δ_1-function $\Gamma : \gamma \to B - \mathcal{L}^2_\gamma$ and $x \in L_\gamma$ we have

$$\forall y \subseteq x \, \exists \alpha \in L_\gamma (L_\gamma \models \varphi_\alpha(y))$$

$$\Downarrow$$

$$\exists \beta < \gamma \, \forall y \subseteq x \, \exists \alpha < \beta \, L_\beta \models \varphi_\alpha(y),$$

then there is no computation in ${}^3E$ with arguments from ω of length γ.

Remarks: Theorem 20 c and d together with theorem 18 gives a new characterization of $\gamma(\omega,0)$ while Theorem 20 a and b strengthen the result from Theorem 19. In both a and b we may let γ be any 2nd order admissible ordinal that is not a cardinal, in c γ may be any 2nd order admissible ordinal and in d we may, by a technically more complicated proof, deduce that γ is 2nd order admissible. Since our main interest is the gaps, we restrict ourselves to the present theorem.

Proof: a By lemma 47 and Δ_1-separation we may apply the proof of Theorem 19 directly.

b Clearly $L_\gamma \subseteq A^\gamma_1$, so $\mathcal{P}(x) \in A^\gamma_2$.

Regard

$$B = \{(X,Y)\,;\, Y \subseteq \omega \text{ codes an initial segment } L_\beta \text{ of } L \text{ and } x \in L_\beta \text{ and } X \text{ codes a subset } x(X) \text{ of } x\}$$

Denote β by $\beta(Y)$ when Y is a code for L_β. We then have

$B \in \Delta_2^\gamma$ and

$$* \quad \forall (X,Y) \in B\ \exists \alpha (\beta(Y) < \alpha \quad (\text{if } \beta(Y) < \gamma)$$
$$\text{or } \exists Z \subseteq Y\ (X,Z) \in B \ \&\ L_{\beta(Z)} \models \varphi_\alpha(x(X))) \quad (\text{if } \beta(Y) \geq \gamma)$$

By lemma 47 the matrix is Δ_1 over $\mathfrak{A}^\gamma$, so let β be a bound for the α's. Let $y \subseteq x$, Y be a code for L_γ and X a code for y. We then see from * that for some $\alpha < \beta$ and some $\gamma_y \leq \gamma$, $L_{\gamma_y} \models \varphi_\alpha$.

But we then have

$$L_\gamma \models \forall y \subseteq x \bigvee_{\alpha<\beta} (\varphi_\alpha(y) \vee \exists \delta\, L_\delta \models \varphi_\alpha(y))$$

This is a formula in $\mathcal{L}^2_\gamma$, and by a, this formula reflects from γ. But that is exactly what we want to prove.

c Note that formulas in $B - \mathcal{L}^2_\gamma$ are absolute. Then c is nothing but a special case of b. (We may give a simpler, direct proof of c.)

d We are going to define formulas φ_α such that

i $\varphi_\alpha \in B - \mathcal{L}^2_\gamma$

ii φ_α is Δ_1-definable from α

iii φ_α is defined on the set of sequences σ from I, and $\varphi_\alpha(\sigma) \Longleftrightarrow \sigma$ is a computation and $|\sigma| < \alpha$.

Construction:

$\varphi_0(\sigma)$: $\sigma \neq \sigma$

$\varphi_\lambda(\sigma) = \bigvee_{\delta<\lambda} \varphi_\delta(\sigma)$ when λ is a limit ordinal

$\varphi_{\alpha+1} = \varphi_\alpha \vee \Gamma(\varphi_\alpha)$ where Γ is the inductive operator defining computations in ${}^3E$.

Now, if ${}^3E(\lambda x\{e\}(x,\vec{n}))$ is a computation of length $\leq \gamma$, then

$$\forall x \in I\ \exists \alpha\ \exists n\ \varphi_\alpha(e,x,\vec{n},n)$$

By the reflection we find a bound β below γ. □

We end the investigation of $\gamma(\omega,0)$ by pointing out some relations between $L_{\gamma(\omega,0)}$ and $L_{\aleph_1}$ under certain set-theoretic assumptions.

Lemma 48: Let γ be strictly countable and the beginning of a gap. Let $\varphi \in \mathcal{L}^1_\lambda$ be a formula with free variables $x_1,\dots,x_n$. Then the following set is recursive in ${}^3E$ by a computation shorter than γ:

$W_\varphi = \{w \in WO ; \varphi \in \mathcal{L}^1_{\|w\|} \ \& \ L_{\|w\|} <_\varphi L_{\aleph_1}\}$

where

$L_\alpha <_\varphi L_\beta$ if $\alpha \le \beta$ and for all subformulas ψ of φ, $L_\alpha <_\psi L_\beta$, and for all $x_1,\dots,x_n \in L_\alpha$

$L_\alpha \models \varphi(x_1,\dots,x_n) \iff L_\beta \models \varphi(x_1,\dots,x_n)$.

Proof: We may assume that φ is built up by $\neg$, $\exists x_i$, W (infinitistic 'or') and atomic formulas. The description of W_φ as a recursive set is by induction on the complexity of φ. We will give a description of $On(\varphi) = \{\alpha < \aleph_1 ; \varphi \in \mathcal{L}^1_\alpha \ \& \ L_\alpha <_\varphi L_{\aleph_1}\}$.

i φ atomic: $On(\varphi)$ consists of those $\alpha < \aleph_1$ such that the parameters in φ are in L_α, i.e. $\varphi \in L_\alpha$. We describe W_φ by using recursive, strictly countable codes for the parameters. Clearly this $On(\varphi)$ has the right properties.

ii $\varphi = \neg\psi$: $\alpha \in On(\varphi) \iff \varphi \in L_\alpha \ \& \ \alpha \in On(\psi)$. It is trivial to see that if $On(\psi)$ has the wanted properties, so will $On(\varphi)$.

iii $\varphi = \underset{y \in x}{W} \psi_y$: $\alpha \in On(\varphi) \iff \forall y \in x \ \alpha \in On(\psi_y)$. Let $\alpha \in On(\varphi)$. If ψ is a subformula of φ, ψ is either ψ_y or a subformula of ψ_y for some $y \in x$. Then $L_\alpha \models \psi(x_1,\dots,x_k) \iff L_{\aleph_1} \models \psi(x_1,\dots,x_k)$ by induction hypothesis.

Assume $L_\alpha \models \varphi(x_1,\dots,x_n)$. Then there is a y such that $L_\alpha \models \psi_y(x_1,\dots,x_n)$. Then $L_{\aleph_1} \models \psi_y(x_1,\dots,x_n)$ and

$L_{\aleph_1} \models \varphi(x_1,\ldots,x_n)$. By the same proof

$L_{\aleph_1} \models \varphi(x_1,\ldots,x_n) \Rightarrow L_\alpha \models \varphi(x_1,\ldots,x_n)$

Now assume $L_\alpha <_\varphi L_{\aleph_1}$. In particular then $L_\alpha <_{\psi_y} L_{\aleph_1}$ for all $y \in x$, so $\alpha \in On(\varphi)$ as defined.

To translate this to a relation in WO recursive in ${}^3E$, use a strictly countable code for x recursive in ${}^3E$ below γ.

iv $\varphi = \exists x \psi$:

$\alpha \in On(\varphi) \Longleftrightarrow \alpha \in On(\psi)\ \&\ \forall\beta(\alpha \leq \beta < \aleph_1\ \&\ \beta \in On(\psi) \Longrightarrow L_\alpha <_\varphi L_\beta)$

We first prove that this is the $On(\varphi)$ we want to define.

Let $\alpha \in On(\varphi)$. If ψ_o is a subformula of φ, ψ_o is either ψ or a subformula of ψ. Thus $L_\alpha <_{\psi_o} L_{\aleph_1}$ since $\alpha \in On(\psi)$.

Assume $L_\alpha \models \exists x\, \psi(x,x_1,\ldots,x_n)$. Since $\alpha \in On(\psi)$, $L_{\aleph_1} \models \exists x\, \psi(x,x_1,\ldots,x_n)$ so $L_{\aleph_1} \models \varphi(x_1,\ldots,x_n)$. Assume $L_{\aleph_1} \models \exists x\, \psi(x,x_1,\ldots,x_n)$, and pick one such x in $L_{\aleph_1}$. Let β_o be such that $x \in L_{\beta_o}$, $\alpha < \beta_o$. Let $\{f_i\}_{i\in\omega}$ be a set of Skolem-functions for ψ and all its subformulas over $L_{\aleph_1}$. Let $\beta_1 < \aleph_1$ be such that the values of the f_i's applied on L_{β_o} is in L_{β_1}. Let $\beta_2 < \aleph_1$ be such that the values of the f_i's applied on L_{β_1} is in L_{β_2} and so on. Let $\beta = \sup\{\beta_i\}$. Then L_β is closed under all the Skolem-functions, and $L_\beta <_\psi L_{\aleph_1}$.

So $\beta \in On(\psi)$ and $\beta > \alpha$. By definition of $On(\varphi)$ and $L_\beta \models \varphi(x_1,\ldots,x_1)$. Then $L_\alpha \models \varphi(x_1,\ldots,x_n)$.

Now assume that $L_\alpha <_\varphi L_{\aleph_1}$. We must prove that $\alpha \in On(\varphi)$. In particular, $L_\alpha <_\psi L_{\aleph_1}$, so $\alpha \in On(\psi)$. Let $\beta > \alpha$ and $\beta \in On(\psi)$. We must prove that $L_\alpha <_\varphi L_\beta$. For subformulas ψ_o this is trivial since $L_\alpha <_\psi L_{\aleph_1}$ and $L_\beta <_\psi L_{\aleph_1}$. Also

$L_\alpha \models \exists x\, \psi(x,x_1,\ldots,x_n) \Rightarrow L_\beta \models \exists x\, \psi(x,x_1,\ldots,x_n)$ is trivial, since $L_\alpha <_\psi L_\beta$.

Assume $L_\beta \models \exists x\, \psi$. Then $L_{\aleph_1} \models \exists x\, \psi$ sunce $\beta \in On(\psi)$. Since $L_\alpha <_\varphi L_{\aleph_1}$ we also have $L_\alpha \models \exists x\, \psi$.

The construction of $On(\varphi)$ involves a quantifier over $\aleph_1$, which in the definition of $WO(\varphi)$ is a quantifier over WO. We may also be able to describe $L_{\|w_1\|} <_\varphi L_{\|w_2\|}$, which is done by induction on the complexity of φ. □

Remark: By the Skolem-Löwenheim argument given in this proof we see that $On(\varphi)$ is cofinal in $\aleph_1$. This is used for the following:

Theorem 21: Assume that recursion in ${}^3E$ satisfies WO-uniformization. Let $\gamma < \aleph_1$ be the beginning of a gap. Then L_γ is an elementary substructure of $L_{\aleph_1}$ with respect to $\mathcal{L}^1_\gamma$.

Proof: It is sufficient to show that $On(\varphi)$ is cofinal in γ for any $\varphi \in \mathcal{L}^1_\gamma$.

Let $\alpha < \gamma$ be such that $\varphi \in L_\alpha$. Let w_0 be recursive below γ such that $\|w_0\| = \alpha$.

Let $X = \{w \in WO(\varphi) ; \|w\| > \|w_0\|$ and $\|w\|$ is minimal for $w \in WO(\varphi)$ with this property$\}$

By WO-uniformization pick $w \in X$ recursive below γ. Then $\gamma > \|w\|$ since there are computations in w of all lengths $\leq \|w\|$. Let $\beta = \|w\|$, $\alpha < \beta < \gamma$ and $\beta \in On(\varphi)$.

Since γ is a closure point of all $On(\varphi)$ (which are closed) we have $\gamma \in \bigcap_{\varphi \in \mathcal{L}^1_\gamma} On(\varphi)$. □

Theorem 22: Let γ be the beginning of a gap.

a If $\aleph_1$ is accessible in L, and we have WO-uniformization or Δ^1_2-determinacy (see Moschovakis [25]) then γ is not a cardinal in L, $(\gamma^+)_L = \aleph_1$ and L_γ is an elementary submodel of $L_{\aleph_1}$.

b If $\aleph_1$ is inaccessible in L, γ is strictly countable and we have WO-uniformization or Δ^1_2-determinacy, then γ is a cardinal of L, and $\gamma = (\aleph_\gamma)_L$.

c Assume that there is a transitive set $X \models ZF$. Then L_γ contains the minimal model for ZFC.

Remark: For any recursively axiomatizable theory Γ in the 1 st order language of set theory, if some β exists such that $L_\beta \models \Gamma$, then for some $\beta < \gamma$ $L_\beta \models \Gamma$. We may use the same proof as in c.

Proof:

a Since $\aleph_1$ is regular, $\aleph_1$ will be a successor cardinal in L. Let α be the last countable ordinal that is a cardinal in L.

Claim: $\{w ; \|w\| = \alpha\}$ is Π^1_3-definable.

Proof: $\|w\|$ is a cardinal in L if for all $A \subseteq \omega \times \omega$ (if A is isomorphic to an initial segment L_β of L and $\beta > \|w\|$ then L_β contains no 1-1-mapping of $\|w\|$ onto a smaller ordinal).

"A is isomorphic to an initial segment L_β of L" is Π^1_1, comparison of two well-founded relations is Δ^1_1, to find the n in field A that is mapped on $\|w\|$ by the isomorphism is Δ^1_1, and "L_β contains no 1-1-mapping of $\|w\|$ onto a smaller ordinal" is 1 st order over L_β and thus arithmetic over A. Thus "$\|w\|$ is a cardinal in L" is a Π^1_2-relation. $\|w\| = \alpha \iff \|w\|$ is a cardinal in L and $\forall w_0 \in WO$

$(\|w_0\| > \|w\| \Rightarrow \|w_0\|$ is not a cardinal in L)

This is seen to be Π^1_3 and the claim is proved.

Moschovakis [25] proved $\mathrm{Det}(\underset{\sim}{\Delta}^1_2) \Rightarrow \mathrm{Uniform}(\Pi^1_3)$ so we may by both assumptions pick one $w \in WO$ that is recursive in ${}^3E$ below γ such that $\|w\| = \alpha$. We have seen before that then $\alpha < \gamma$. This proves the first part of a. To see the second part, let w be recursive in ${}^3E$ as above. We will prove that above $\|w\|$ we have WO-uniformisation. So let $\beta > \alpha$ be given and assume that $\{w_0 \in WO; \|w_0\| = \beta\}$ is recursive in ${}^3E$.

Since $(\mathrm{card}(\beta))_L = \alpha$ there will be a constructible subset X of $\alpha \times \alpha$ being a well-ordering of length β. There is a canonical isomorphism φ between $\alpha \times \alpha$ and field(w) $\times$ field(w), and by standard methods we see that

$$Q = \{Y \subseteq \text{field}(w) \times \text{field}(w) : \varphi''y \text{ is constructible}\}$$

is Σ_2^1, and moreover, Q has a Σ_2^1-well-ordering.
Now we pick the minimal element of Q such that $\varphi''(y)$ is a well-ordering of length β. The characteristic function x of y is an element of A, and x is by these considerations effectively selected from A. Thus x is recursive in ^3E by a computation not much longer than that of A.

Remark: From this proof we may extract a simple argument for the following:

If $\aleph_1 = (\aleph_1)_L$, then we have WO-uniformisation.

b Let $w_0 \in WO$ be given. $\|w_0\| = \alpha < \gamma$.
By an argument like that of a we prove that $\{w ; \|w\| = (\alpha^+)_L\}$ is $\Pi_3^1(w_0)$ and we may select one w_1 from this set. Then $w_1 < \gamma$. We may iterate this process transfinitely and the iteration will stop when the length of the iteration is not recursive below γ, i.e. $(\alpha^{(\beta)})_L = \alpha^{\overbrace{++\cdots+}^{\beta}}{}_L < \gamma$ as long as $\beta < \gamma$. This proves b.

c By the assumption there will be some ordinal β such that $L_\beta \models ZFC$.
Since Skolem-Löwenheim arguments are valid inside L there will be a β_0 countable in L such that $L_{\beta_0} \models ZFC$.
Since $\{w ; \|w\| = \beta_0\}$ contains constructible elements, we may select the least constructible $w \in WO$ such that $\|w\| = \beta_0$. w is recursive in ^3E by a short computation, so $\beta_0 < \gamma$. □

§ 12 ON PLATEK: "FOUNDATIONS OF RECURSION THEORY"

Is it possible to define a recursion theory without introducing indices? In § 2 we defined an operator Γ which generates computations rather than functions. The partial recursive function φ with index e is defined by: $\varphi(\text{——}) \simeq y$ iff $\langle e, \text{——}, y\rangle \in \Gamma^{\infty}$. Because of the scheme of diagonalization (scheme XIII) it is not quite obvious how to find an indexfree procedure which gives the partial recursive functions. There is such a procedure for the primitive recursive functions: They are the least set of functions containing the initial functions and closed under composition and primitive recursion. However in Platek: "Foundations of Recursion Theory" ([26]) there is an indexfree approach to the partial recursive functions. In this chapter we prove that his system is equivalent to recursion theory with indeces and schemes, as in § 2. It also turns out that theorem 5.3.2 in [26] (First Recursion Theorem) can easily be proved from the results which lead up to this equivalence. In [26] the theorem is proved by λ-calculus.

As [26] has not been published some of the notions, definitions and results from [26] will be given here, enough to make this chapter selfcontained.

Type symbols, total types, partial types, hereditarily consistent types.

The type symbols, denoted by TS, is the least set of symbols such that $0 \in TS$, if $\sigma \in TS$, $\tau \in TS$ then $\sigma \to \tau \in TS$.

If $\sigma \in TS$, $\sigma \neq 0$, then there are unique type symbols $\sigma_1 \dots \sigma_k$ such that $\sigma = \sigma_1 \to (\sigma_2 \to \dots \to (\sigma_k \to 0)\dots)$. $\sigma_1 \dots \sigma_k$ are called the factors of σ. The number $l(\sigma)$ (the level of σ) is defined by: $l(0) = 0$, if $\sigma \neq 0$ then $l(\sigma) = \max\{l(\sigma_i) + 1 : \sigma_i \text{ is a factor of } \sigma\}$. The pure type symbols are the following ones: 0 is a pure type symbol. If σ is a pure type symbol then $\sigma \to 0$ is a pure type symbol. The pure type symbols are also denoted by $0, 1, 2, \dots, k, \dots$, where $k+1 =$

$k \to 0$, $k \geq 0$. σ is a <u>special</u> type symbol if all its factors are pure. For any type symbol σ there is a special type symbol $Sp(\sigma)$ defined by: $Sp(0) = 0$, $Sp(\sigma \to \tau) = l(\sigma) \to Sp(\tau)$. Then $Sp(\sigma)$ is special, $l(\sigma) = l(Sp(\sigma))$, if σ is special then $Sp(\sigma) = \sigma$.

For each type symbol σ we associate a number $c(\sigma)$ (the <u>code</u> of σ) in the following way: $c(0) = 1$, $c(\sigma \to \tau) = 2^{c(\sigma)} \cdot 3^{c(\tau)}$. c is a one-to-one function from TS to the natural numbers. The <u>signature</u> of a list $\sigma_1 \ldots \sigma_k$ of type symbols is the number $\langle c(\sigma_1), \ldots, c(\sigma_k) \rangle$, where $\langle x_1 \ldots x_k \rangle = p_1^{x_1} \cdot p_2^{x_2} \ldots p_k^{x_k}$, p_i is the i-th prime number. Different sequences of type symbols will have different signatures.

Sometimes we omit the parantheses and write $\sigma_1 \to \sigma_2 \to \sigma_3 \to \ldots \to \sigma_k$ instead of $\sigma_1 \to (\sigma_2 \to \ldots\ldots \to (\sigma_{k-1} \to \sigma_k) \ldots)$.

Let Ob be any set. For each $\sigma \in TS$ we define a set $T(\sigma)$ (= the set of <u>total objects</u> of type σ), a set $PT(\sigma)$ (= the set of <u>partial objects</u> of type σ), and a set $HC(\sigma)$ (= the set of <u>hereditarily consistent objects</u> of type σ). If $\sigma \neq 0$ we also define a partial relation $\leq$ on $HC(\sigma)$.

$T(0) = Ob$. If $\sigma \neq 0$ let $T(\sigma) = T(\sigma_1) \times T(\sigma_2) \times \ldots \times T(\sigma_k) \to Ob$ (= the set of functions defined on $T(\sigma_1) \times \ldots \times T(\sigma_k)$ and with values in Ob), where $\sigma_1 \ldots \sigma_k$ are the factors of σ.

$PT(0) = Ob$. If $\sigma \neq 0$ let $PT(\sigma) = T(\sigma_1) \times T(\sigma_2) \times \ldots \times T(\sigma_k) \overset{p}{\to} Ob$ (= the set of functions defined on a subset of $T(\sigma_1) \times \ldots \times T(\sigma_k)$ and with values in Ob.

$HC(0) = Ob$. If $\sigma \neq 0$ let $HC(\sigma)$ be the set of partial monotone functions defined on a subset of $HC(\sigma_1) \times HC(\sigma_2) \times \ldots \times HC(\sigma_k)$, and with values in Ob. By "monotone" we mean: if $f(g^{\sigma_1}, \ldots, g^{\sigma_k}) \simeq x$ and $g^{\sigma_1} \leq h^{\sigma_1}, \ldots, g^{\sigma_k} \leq h^{\sigma_k}$, then $f(h^{\sigma_1}, \ldots, h^{\sigma_k}) \simeq x$. If $f \in HC(\sigma)$, $f' \in HC(\sigma)$ then $f \leq f'$ if f' is an extension of f.

We easily see that $PT(\sigma) = HC(\sigma)$ if $l(\sigma) \leq 1$, $T(\sigma) \subseteq PT(\sigma)$ for any σ. Note that $T(\sigma \to \tau)$ can be identified with $T(\sigma) \to T(\tau)$ (= the set of functions defined on $T(\sigma)$ and with values in $T(\tau)$. If

$\tau \neq 0$ then $PT(\sigma \to \tau)$ can be identified with $T(\sigma) \to PT(\tau)$. If $\tau \neq 0$ then $HC(\sigma \to \tau)$ can be identified with the set of monotone functions defined on $HC(\sigma)$ and with values in $HC(\tau)$. ($f \in HC(\sigma) \to HC(\tau)$ is monotone if for all $g, g' \in HC(\sigma)$: $g \leq g' \implies f(g) \leq f(g')$.) Suppose $\sigma = \sigma_1 \to \sigma_2 \to \ldots \to \sigma_k \to \tau$. Sometimes we will write $f^{\sigma} g^{\sigma_1} \ldots g^{\sigma_k}$ for $f^{\sigma}(g^{\sigma_1})(g^{\sigma_2}) \ldots (g^{\sigma_k})$. The superscript σ in f^{σ} denotes the type of f. Let $HC = \bigcup \{HC(\tau) : \tau \in TS\}$.

For each $\sigma \in TS$ there is a natural injection $\Psi : PT(\sigma) \to HC(\sigma)$, and a surjection $\Phi : HC(\sigma) \to PT(\sigma)$, such that $\Phi(\Psi(\varphi)) = \varphi$ when $\varphi \in PT(\sigma)$; if φ is a subfunction of φ', $\varphi, \varphi' \in PT(\sigma)$, then $\Psi\varphi \leq \Psi\varphi'$; if $f, f' \in HC(\sigma)$, $f \leq f'$ then Φf is a subfunction of $\Phi f'$.

Ψ and Φ are defined as follows: If $\sigma = 0$ then $\Psi(x) = \Phi(x) = x$. Suppose $\sigma \neq 0$. Let $\sigma_1 \ldots \sigma_k$ be the factors of σ. If $\varphi \in PT(\sigma)$ let $\Psi(\varphi)(g^{\sigma_1} \ldots g^{\sigma_k}) \simeq \varphi(\Phi(g^{\sigma_1}), \ldots, \Phi(g^{\sigma_k}))$, where the right hand side is undefined if $\Phi(g^{\sigma_i})$ is not total for some i. If $f \in HC(\sigma)$ let $\Phi(f)(\beta^{\sigma_1} \ldots \beta^{\sigma_k}) \simeq f(\Psi(\beta^{\sigma_1}), \ldots, \Psi(\beta^{\sigma_k}))$.

If $f \in HC(\sigma)$ then $\Psi(\Phi(f)) \leq f$. In general $\Psi(\Phi(f)) \neq f$. If $l(\sigma) \leq 1$ then $PT(\sigma) = HC(\sigma)$, and $\Psi(\varphi) = \Phi(\varphi) = \varphi$ for all $\varphi \in PT(\sigma)$.

Now we put some conditions on the set Ob. Suppose there is a copy N of natural numbers such that $N \subseteq Ob$. Suppose that there are functions M, K, L such that M is a pairing function $Ob \times Ob \to Ob$ with inverse functions K and L (i.e. $K(M(x,y)) = x$, $L(M(x,y)) = y$ for all $x, y \in Ob$). Later, when we prove the equivalence between recursion theories with and without indeces we will let Ob the universe of a computation domain, so these conditions will be satisfied. We will define translation functions from $HC(\sigma)$ to $HC(\tau)$ for some types σ, τ, and pairing functions on $HC(\sigma)$ for each σ. The following functions will be defined:

$$u_j \in HC(j \to j+1)$$

$$d_j \in HC(j+1 \to 1)$$

$$u_j^{j+k} \in HC(j \to j+k)$$
$$d_j^{j+k} \in HC(j+k \to j)$$
$$M_k^s \in HC(\underbrace{s \to s \to \ldots\ldots \to s}_{k} \to s), \quad k \geq 0$$
$$K_{k,j}^s \in HC(s \to s), \quad 1 \leq j \leq k .$$

If $\tau \in TS$, $Sp(\tau) = \sigma$, $l(\tau) = m$ we define

$$Tr_\mu^\nu \in HC(\mu \to \nu)$$
$$Intr_\mu^\nu \in HC(\nu \to \mu)$$

when (μ,ν) is one of the pairs (m,σ), (σ,m), (σ,τ), (τ,σ), (τ,m), (m,τ). "Tr" stands for translation, "Intr" for inverse translation. Finally we define

$$M^m : \bigcup_{k<\omega} (HC^m)^k \to HC(m)$$
$$K_j^m : HC(m) \to HC^m$$
$$lh^m : HC(m) \to N$$

the predicate $Seq^m \subseteq HC(m)$

where $HC^m = \bigcup \{HC(\tau) : l(\tau) \leq m\}$.

These functions will have the following properties: $d_j(u_j(f^j)) = f^j$, $d_j^{j+k}(u_j^{j+k}(f^j)) = f^j$, $K_{k,j}^s(M_k^s(f_1^s \ldots f_k^s)) = f_j^s$ if $1 \leq j \leq k$, $Intr_\mu^\nu(Tr_\mu^\nu(f^\mu)) = f^\mu$, $K_j^m(M^m(f^{\tau_1}, f^{\tau_2} \ldots f^{\tau_k})) = f^{\tau_j}$ if $1 \leq j \leq k$, $l(\tau_1) \leq m \ldots l(\tau_k) \leq m$, $lh^m(M^m(f^{\tau_1} \ldots f^{\tau_k})) = k$.

The definitions are as follows:

$$u_o(x) = f^1, \text{ where } f^1(y) \simeq x \text{ for all } y$$
$$d_o(f^1) \simeq f^1(0)$$
$$u_{j+1}(f^{j+1})(g^{j+1}) \simeq f^{j+1}(d_j(g^{j+1}))$$
$$d_{j+1}(f^{j+2})(g^j) \simeq f^{j+2}(u_j(g^j))$$
$$u_j^j(f^j) = d_j^j(f^j) = f^j$$

$$u_j^{j+k+1}(f^j) = u_{j+k}(u_j^{j+k}(f^j))$$

$$d_j^{j+k+1}(f^{j+k+1}) = d_j^{j+k}(d_{j+k}(f^{j+k+1}))$$

$$M_o^o = 0\ ,\quad M_1^o(x) = x\ ,\quad M_2^o = M$$

$$M_{k+3}^o(x_1 \ldots x_{k+3}) = M(x_1, M_{k+2}^o(x_2 \ldots x_{k+3}))$$

$$K_{1,1}^o(x) = x\ ,\quad K_{2,1}^o = K\ ,\quad K_{2,2}^o = L$$

$$K_{k+3,1}^o = K\ ,\quad K_{k+3,j+2}^o(x) = K_{k+2,j+1}^o(L(x)) \quad \text{if} \quad 0 \le j \le k+1$$

$$M_o^{s+1} = f_o^{s+1}\ , \quad \text{where} \quad f_o^{s+1}(g^s) \simeq 0\ ,\quad M_1^{s+1}(f) = f$$

$$M_2^{s+1}(f,g)(h^s) \simeq \begin{cases} f(K_{2,2}^s(h)) & \text{if } d_o^s(K_{2,1}^s(h)) \simeq 0 \\ g(K_{2,2}^s(h)) & \text{if } d_o^s(K_{2,1}^s(h)) \simeq y,\ y \neq 0 \\ \uparrow & \text{otherwise} \end{cases}$$

(For total types we could have defined $M_2^1(\alpha,\beta)(x) = M(\alpha(x),\beta(x))$. If we had defined $M_2^1(f,g)(x) \simeq M(f(x),g(x))$ then $M_2^1(f,g)(x)$ would be undefined if one of $f(x), g(x)$ is undefined. Hence we could not always get back f and g from $M_2^1(f,g)$. This is possible with the definition above.)

$$M_{k+3}^{s+1}(f_1 \ldots f_{k+3}) = M_2^{s+1}(f_1, M_{k+2}^{s+1}(f_2 \ldots f_{k+3}))$$

$$K_{1,1}^{s+1}(f) = f\ ,\quad K_{2,1}^{s+1}(f)(h) \simeq f(M_2^s(u_o^s(0),h^s))$$

$$K_{2,2}^{s+1}(f)(h^s) \simeq f(M_2^s(u_o^s(1),h^s))$$

$$K_{k+3,1}^{s+1} = K_{2,1}^{s+1}\ ,\quad K_{k+3,j+2}^{s+1}(f) = K_{k+2,j+1}^{s+1}(K_{2,2}^{s+1}(f))$$

If $\tau = 0$ then $\sigma = Sp(\tau) = 0$, $m = l(\tau) = 0$, and we let $Tr_\mu^\nu(x) = Intr_\mu^\nu(x) = x$. Suppose $\tau \neq 0$. Let $\tau_1, \tau_2, \ldots, \tau_k$ be the factors of τ. Let $l(\tau_i) = n_i$, $i = 1 \ldots k$. Let $\sigma = Sp(\tau)$. Then the factors of σ are $n_1 \ldots n_k$. Let $m = l(\tau)$. Then $m = \max\{n_i + 1 : i = 1 \ldots k\}$. Let i be the least number such that $m = n_i + 1$. Let

$$Tr_m^\sigma(f^m)(g^{n_1} \ldots g^{n_k}) \simeq f^m(g^{n_i})$$

$$Intr_m^\sigma(f^\sigma)(g^{m-1}) \simeq f^\sigma(d_{n_1}^{m-1}g, d_{n_2}^{m-1}g, \ldots, d_{n_k}^{m-1}g)$$

(note that $d^{m-1}_{n_i}(g) = g$ as $m-1 = n_i$)

$$Tr^m_\sigma(f^\sigma)(g^{m-1}) \simeq f^\sigma(d^{m-1}_{n_1}(K^{m-1}_{k,1}(g)),\ldots,d^{m-1}_{n_k}(K^{m-1}_{k,k}(g)))$$

$$Intr^m_\sigma(f^m)(g^{n_1}\ldots g^{n_k}) \simeq f^m(M^{m-1}_k(u^{m-1}_{n_1}(g^{n_1}),\ldots,u^{m-1}_{n_k}(g^{n_k})))$$

$$Tr^\tau_\sigma(f^\sigma)(g^{\tau_1}\ldots g^{\tau_k}) \simeq f^\sigma(Intr^{\tau_1}_{n_1}(g^{\tau_1}),\ldots,Intr^{\tau_k}_{n_k}(g^{\tau_k}))$$

$$Intr^\tau_\sigma(f^\tau)(g^{n_1}\ldots g^{n_k}) \simeq f^\tau(Tr^{\tau_1}_{n_1}(g^{n_1}),\ldots,Tr^{\tau_k}_{n_k}(g^{n_k}))$$

$$Tr^\sigma_\tau(f^\tau)(g^{n_1}\ldots g^{n_k}) \simeq f^\tau(Intr^{n_1}_{\tau_1}(g^{n_1}),\ldots,Intr^{n_k}_{\tau_k}(g^{n_k}))$$

$$Intr^\sigma_\tau(f^\sigma)(g^{\tau_1}\ldots g^{\tau_k}) \simeq f^\sigma(Tr^{n_1}_{\tau_1}(g^{\tau_1}),\ldots,Intr^{n_k}_{\tau_k}(g^{\tau_k}))$$

$$Tr^m_\tau(f^\tau) = Tr^m_\sigma(Tr^\sigma_\tau(f^\tau))$$

$$Intr^m_\tau(f^m) = Intr^\sigma_\tau(Intr^m_\sigma(f^m))$$

$$Tr^\tau_m(f^m) = Tr^\tau_\sigma(Tr^\sigma_m(f^m))$$

$$Intr^\tau_m(f^\tau) = Intr^\sigma_m(Intr^\tau_\sigma(f^\tau))$$

If $\sigma = m$ then it follows from these definitions that $Tr^\sigma_m(f) = Intr^\sigma_m(f) = Tr^m_\sigma(f) = Intr^m_\sigma(f) = f$. If $\tau = \sigma$ it follows that $Tr^\tau_\sigma(f) = Intr^\tau_\sigma(f) = Tr^\sigma_\tau(f) = Intr^\sigma_\tau(f) = f$. In general $Tr^\nu_\mu(Tr^\mu_\nu(f^\nu)) \neq f^\nu$. Tr^ν_μ is an injection $HC(\mu) \to HC(\nu)$, and $Intr^\nu_\mu$ is a surjection $HC(\nu) \to HC(\mu)$. In general $Tr^\nu_\mu(Intr^\nu_\mu(f^\nu)) \neq f^\nu$.

Suppose $l(\tau_i) = m_i \leq m$, $i = 1,\ldots,k$. Let $M^m(f^{\tau_1}\ldots f^{\tau_k}) = M^m_{k+1}(u^m_o(p),u^m_{m_1}(Tr^{m_1}_{\tau_1}(f^{\tau_1})),\ldots,u^m_{m_k}(Tr^{m_k}_{\tau_k}(f^{\tau_k})))$, where p is the signature of the list $\tau_1,\ldots,\tau_k$. Let

$K^m_j(f^m) = Intr^{m_j}_{\tau_j}(d^m_{m_j}(K^m_{k+1,j+1}(f^m)))$ if $1 \leq j \leq k$, where the numbers k, m_j and the types τ_i can be found from $d^m_o(K^m_{2,1}(f^m))$. Let $K^m_j(f^n) = f$ if $j = 0$ or $j > k$. Let $l^m(f^m) = k$ where $d^m_o(K^m_{2,1}(f^m)) = p_1^{x_1},p_2^{x_2},\ldots,p_k^{x_k}$, $p_1,p_2,\ldots$ are the prime numbers in increasing order.

Let

$$\mathrm{Seq}^m = \{f^m : \exists f^{\tau_1} \ldots f^{\tau_k},\ l(\tau_i) \leq m \text{ for } i = 1 \ldots k,\ f^m = M^m(f^{\tau_1} \ldots f^{\tau_k})\}.$$

Recursion on a computation domain $\mathfrak{A}$.

Let $\mathfrak{A} = (A,N,s,M,K,L,O)$, where N is the set of natural numbers, A is any set such that $N \subseteq A$, $s(x) = x+1$ if $x \in N$, $s(x) = 0$ if $x \notin N$, M is a pairing function on A with inverse functions K, L. We will consider five recursion theories on this domain, four theories with indeces and one without. Let $Ob = A$.

First we regard recursion on $\mathfrak{A}$ relative to a list $\varphi_1,\ldots,\varphi_l$, $\mathcal{F}_1,\ldots,\mathcal{F}_m$ where $\varphi_i \in PT(\sigma)$ with $l(\sigma) = 1$, $\mathcal{F}_i$ is a monotone partial functional, i.e. $\mathcal{F}_i$ is a monotone partial function defined on a subset of $PT(\sigma_1) \times PT(\sigma_2) \times \ldots \times PT(\sigma_j)$ and with values in A, where $l(\sigma_t) \leq 1$, $t = 1,\ldots,j$. $\mathcal{F}_i$ can be regarded as an object in $HC(\tau)$ with $l(\tau) = 2$ because $PT(\sigma_t) = HC(\sigma_t)$ when $l(\sigma_t) \leq 1$. Below we give the schemes to the left and the indeces to the right. —— is a list of objects from A. $\psi_1,\ldots,\psi_p$ are partial functions with indeces $e_1,\ldots,e_p$. ψ has index e. ——' is the list obtained from —— by moving the j-th element in —— to the front of the list. n is the signature of the argument list of φ.

S1	$\varphi(x, \text{——})$	$= \begin{cases} 0 & \text{if } x \in N \\ 1 & \text{if } x \notin N \end{cases}$	$\langle 1,n \rangle$
S2	$\varphi(x, \text{——})$	$= x$	$\langle 2,n \rangle$
S3	$\varphi(x,y, \text{——})$	$= \begin{cases} 0 & \text{if } x = y \\ 1 & \text{if } x \neq y \end{cases}$	$\langle 3,n \rangle$
S4	$\varphi(\text{——})$	$= 0$	$\langle 4,n \rangle$
S5	$\varphi(x, \text{——})$	$= \begin{cases} x+1 & \text{if } x \in N \\ 0 & \text{if } x \notin N \end{cases}$	$\langle 5,n \rangle$

S6	$\varphi(x,y,$ ——$) = M(x,y)$		$\langle 6,n\rangle$
S7	$\varphi(x,$ ——$) = K(x)$		$\langle 7,n\rangle$
S8	$\varphi(x,$ ——$) = L(x)$		$\langle 8,n\rangle$
S9	$\varphi($ —— $) \simeq \psi_1(\psi_2($—$),$—$)$		$\langle 9,n,e_1,e_2\rangle$
S10	$\varphi(0,$ ——$) \simeq \psi_1($——$)$		$\langle 10,n,e_1,e_2\rangle$
	$\varphi(k+1,$ ——$) \simeq \psi_2(\varphi(k,$——$),k,$——$)$		
	$\varphi(x,$ ——$) = 0 \quad \text{fi} \quad x \notin N$		
S11	$\varphi($ —— $) \simeq \psi($——$')$		$\langle 11,n,e,j\rangle$
S12	$\varphi(x,$ ——$) \simeq \{x\}($——$)$		$\langle 12,n\rangle$
S13	$\varphi(x_1 \ldots x_k,$—$) \simeq \varphi_i(x_1 \ldots x_k)$		$\langle 13,n,i\rangle$
S14	$\varphi($——$) \simeq \mathcal{F}_i(\lambda\vec{x}\,\psi_1(\vec{x},$—$),\ldots,\lambda\vec{x}\,\psi_k(\vec{x},$—$))$		$\langle 14,n,i,\langle e_1 \ldots e_k\rangle\rangle$

This recursion theory will be denoted by I . $\{e\}^{I}$ denotes the partial function with index e in I .

In the second system we will allow partial functions with total objects of pure type as arguments. Let —— be a list of objects of pure type. S1 - S14 will be as in I . The number n at the second place in the index is the signature of the argument list. In S11 —— is the list obtained from —— by moving the j-th object in —— to the front of the list. The index is $\langle 11,n',e,j\rangle$, where n' is the signature for the list ——' . In addition there are two new schemes.

S15	$\varphi(x,\alpha^1,$——$) = \alpha\,(x)$	$\langle 15,n\rangle$
S16	$\varphi(\alpha^{k+2},$——$) \simeq \alpha^{k+2}(\lambda\beta^k\psi(\beta,\alpha,$——$))$	$\langle 16,n,\langle e\rangle\rangle$

This recursion theory will be denoted by II . $\{e\}^{II}$ denotes the partial function with index e .

In the third system we allow partial functions with total objects of any type as arguments. Now —— is a list of total objects of any type. The number n in the index is the signature of the argument list of φ. S1 - S15 are as in II. S16 is changed as follows:

$$\text{S16} \quad \varphi(\alpha^{\tau}, \text{——}) \simeq \alpha^{\tau}(\delta_1,\ldots,\delta_k) \qquad \langle 16,n,\langle e_1 \ldots e_k\rangle\rangle$$

where $\tau \neq 0$, $\delta_i = \lambda\vec{\gamma}\,\psi_i(\vec{\gamma}, \alpha^{\tau}, \text{——})$, and the type of δ_i is the same as the i-th factor of τ. (We allow $\vec{\gamma}$ to be the empty sequence for the case where the i-th factor of τ is 0.)

This system will be denoted by III, and $\{e\}^{III}$ denotes the partial function with index e.

In the fourth system we allow partial functions with hereditarily consistent objects as arguments. S1 - S16 are as in III, with each variable for total objects of type σ substituted by a variable for hereditarily consistent objects of type σ. In S15 "$=$" is substituted by "$\simeq$". This system will be denoted by IV, and $\{e\}^{IV}$ denotes the partial function with index e.

Now S16 has the form:

$\varphi(f^{\tau}, \text{——}) \simeq f^{\tau}(p_1,\ldots,p_k)$, where $p_i = \lambda\vec{q}\,\psi_i(\vec{q}, f^{\tau}, \text{——})$.
This is to be understood as follows: If $p_i \in HC$, $i = 1,\ldots,k$, and $f^{\tau}(p_1,\ldots,p_k) \simeq y$ then $\varphi(f^{\tau}, \text{——}) \simeq y$. Let e be the index for φ, $e_1,\ldots,e_n$ indeces for $\psi_1,\ldots,\psi_k$. The length of the computation $\{e\}^{IV}(f^{\tau}, \text{——})$ is the least ordinal α such that $(p_i)_{\alpha} \in HC$ for $i = 1,\ldots,k$, and $f^{\tau}((p_1)_{\alpha},\ldots,(p_k)_{\alpha})$ is defined. If the type of p_i is not 0 then $(p_i)_{\alpha}$ is defined by: $(p_i)_{\alpha}(\vec{q}) \simeq \{e_i\}^{IV}(\vec{q}, f^{\tau}, \text{——})$ if $|\{e_i\}^{IV}(\vec{q}, f^{\tau}, \text{——})| < \alpha$, $(p_i)_{\alpha}(\vec{q})$ is undefined otherwise. If the type of p_i is 0 then $(p_i)_{\alpha} = \{e_i\}^{IV}(f^{\tau}, \text{——})$ if $|\{e_i\}^{IV}(f^{\tau}, \text{——})| < \alpha$. $(p_i)_{\alpha}$ is $\uparrow$ otherwise. (Convention: $f^{\tau}(\ldots\uparrow\ldots)$ is undefined.) We let $|\{e\}^{IV}(f^{\tau}, \text{——})|$ denote the length of the computation $\{e\}^{IV}(f^{\tau}, \text{——})$.

The fifth system is Platek's indexfree recursion theory from [26].

Suppose $\mathcal{B} \subseteq HC$. $\mathcal{R}_\omega(\mathcal{B})$ is the least subset X of $HC \cup \{\uparrow\}$ such that $\mathcal{B} \subseteq X$, $DC \in X$, all $I, K, S, FP \in X$, and if $f^{\sigma\to\tau} \in X$, $g^\sigma \in X$ then $fg \in X$. ($\uparrow$ is "being undefined".)

Let $k \in N$. $\mathcal{R}_k(\mathcal{B})$ is defined as $\mathcal{R}_\omega(\mathcal{B})$ except that $FP^{(\tau\to\tau)\to\tau}$ is an initial function only when $l(\tau) \leq k$.

The objects DC, I, K, S, FP are defined as follows. $DC \in HC(0\to 0\to 1\to 1)$.

$$DC(x,y,f^1,g^1) = \begin{cases} f & \text{if } x = y \\ g & \text{if } x \neq y \end{cases}$$

For each σ $I^{\sigma\to\sigma} \in HC(\sigma\to\sigma)$. For each σ,τ $K^{\sigma\to\tau\to\sigma} \in HC(\sigma\to\tau\to\sigma)$, for each σ,τ,λ $S^\mu \in HC(\mu)$, where $\mu = (\sigma\to\tau\to\lambda)\to(\sigma\to\tau)\to\sigma\to\lambda$.

$$I(f^\sigma) = f^\sigma$$

$$K(f^\sigma, g^\tau) = f^\sigma$$

$$S(f^{\sigma\to\tau\to\lambda}, g^{\sigma\to\tau}, h^\sigma) = f(h)(g(h))$$

(if $\lambda = 0$ we substitute "$=$" by "$\simeq$" in the definition of S.)

If $f \in HC(\tau\to\tau)$, $\tau \neq 0$, then there is a least object $g \in HC(\tau)$ such that $f(g) = g$. This can be proved as follows: Let $g_0 = \uparrow^\tau$ (= the hereditarily consistent object of type τ which is totally undefined). Let $g_{\alpha+1} = f(g_\alpha)$, $g_\lambda = \bigcup_{\alpha<\lambda} g_\alpha$ if λ is a limit ordinal. Then $g_\alpha \leq g_\beta$ when $\alpha < \beta$, because $f \in HC$. Let $g = \bigcup_\alpha g_\alpha$, where the union is taken over all ordinals α. Then $g = g_\alpha$ for some α. $f(g) = f(g_\alpha) = g_{\alpha+1} = g$. So g is a fixed point for f. If h is another fixed point for f (i.e. $f(h) = h$) then $g \leq h$. For obviously $g_0 \leq h$, if $g_\alpha \leq h$ then $g_{\alpha+1} = f(g_\alpha) \leq f(h) = h$, hence $g_{\alpha+1} \leq h$.

Let $FP^{(\tau\to\tau)\to\tau}$ be defined as follows: $FP(f^{\tau\to\tau})$ = the least fixed point of f. $FP \in HC((\tau\to\tau)\to\tau)$, for suppose $f_1 \leq f_2$. In order to prove that $FP(f_1) \leq FP(f_2)$ let $g = FP(f_1)$, $h = FP(f_2)$. Obviously $g_0 \leq h_0$. Suppose $g_\alpha \leq h_\alpha$. $g_{\alpha+1} = f_1(g_\alpha) \leq f_1(h_\alpha) \leq f_2(h_\alpha) = h_{\alpha+1}$. This proves that $g \leq h$.

Typed terms. The combinators I, K, S.

To motivate the introduction of I, K, S we prove lemma 49. First we define the typed terms. For each type symbol σ the variables $v_0^\sigma, v_1^\sigma, v_2^\sigma, \ldots$ are terms of type σ. If t_1, t_2 are terms of type $\sigma \to \tau$, σ respectively, then $t_1(t_2)$ is a term of type τ. If t has type τ then $\lambda v_j^\sigma t$ is a term of type $\sigma \to \tau$. The last operation binds the variable v_j^σ. By a closed term we mean a term without free variables.

Given an assignment $f_j^\sigma \in HC(\sigma)$ to v_j^σ for each j, σ, we can extend the assignment to each typed term in the following way: The assignment to $t_1(t_2)$ is $g_1^{\sigma\to\tau}(g_2^\sigma)$, where $g_1^{\sigma\to\tau}$ is the assignment to t_1, g_2^σ is the assignment to t_2. The assignment to $\lambda v_j^\sigma t$ is $g^{\sigma\to\tau}$, where g is defined by: $g(h^\sigma)$ = the assignment to t given that the assignment to v_j^σ is h^σ. Here we allow $\uparrow$ (= being undefined) to be a possible assignment (for the case $\tau = 0$, $g_1^{\sigma\to 0}(g_2^\sigma)$ is undefined). We make the convention that $f^{\sigma\to\tau}(\uparrow) = \uparrow^\tau$ for all $f \in HC(\sigma\to\tau)$, where $\uparrow^\tau \in HC(\tau)$ is the totally undefined object if $\tau \neq 0$, $\uparrow^0 = \uparrow$. It follows that the assignment to a term t of type τ is a hereditarily consistent object ot type τ if $\tau \neq 0$, if $\tau = 0$ the assignment is either an element in $HC(0)$, or $\uparrow$. The assignment to a closed term is independent of the assignments to the variables v_j^σ. These assignments are called explicit definitions (cf. p.20 in [26]).

A combination is a hereditarily consistent object built up from the objects I, K, S by typed function application. Examples of combinations: $K^{\sigma\to\tau\to\sigma}$, $S^\mu K^{\sigma\to\tau\to\sigma}$ where $\mu = (\sigma\to\tau\to\sigma) \to (\sigma\to\tau) \to \sigma \to \sigma$, $S^{\tau_1}(S^{\tau_2}K^{\tau_3}I^{\tau_4})(K^{\tau_5}K^{\tau_6})$ where the types are as follows: Let σ, τ, λ be any type symbols. Let $\tau_6 = \sigma \to \tau \to \sigma$, $\tau_5 = \tau_6 \to (\tau_6 \to \lambda) \to \tau_6$,

$\tau_7 = (\tau_6\to\lambda) \to \tau_6$ (τ_7 is the type of $K^{\tau_5}K^{\tau_6}$),

$\tau_8 = (\tau_6\to\lambda) \to \tau_6 \to \lambda$ (τ_8 is the type of $S^{\tau_2}K^{\tau_3}I^{\tau_4}$),

$\tau_1 = \tau_8 \to \tau_7 \to (\tau_6\to\lambda) \to \lambda$, $\quad \tau_4 = (\tau_6\to\lambda) \to \tau_6 \to \lambda$

$\tau_3 = (\tau_6\to\lambda) \to (\tau_6\to\lambda) \to \tau_6 \to \lambda$, $\tau_2 = \tau_3 \to \tau_4 \to (\tau_6\to\lambda) \to \tau_6 \to \lambda$.

Lemma 49: Let $f \in HC$. Then f is an explicit definition iff f is a combination. (The corresponding result for total types is mentioned in [26] p. 21.)

Proof: Suppose f is combination. If $f = I^{\sigma\to\sigma}$, $K^{\sigma\to\tau\to\sigma}$, $S^{\lambda\to\mu\to\nu\to\rho}$ $(\mu = \nu\to\tau,\ \lambda = \nu\to\tau\to\rho)$ then f is the assignment to the terms $\lambda v^{\sigma}.v^{\sigma}$, $\lambda v^{\sigma}\lambda v^{\tau}.v^{\sigma}$, $\lambda v^{\lambda}\lambda v^{\mu}\lambda v^{\nu}.\ v^{\lambda}(v^{\nu})(v^{\mu}(v^{\nu}))$ respectively. Suppose $f = f_1(f_2)$, where f_1 and f_2 are combinations, and f_1, f_2 are explicit definitions. Let t_1, t_2 be closed terms such that f_1, f_2 are the assignments to t_1, t_2 respectively. Then f is the assignment to $t_1(t_2)$.

Next we prove that the assignment to each closed term is a combination. This will be done by induction on the height of the term, where the height is defined as follows: $h(v^{\sigma}) = 0$, $h(t_1(t_2)) = \max\{h(t_1), h(t_2)\} + 1$, $h(\lambda v^{\sigma} t) = h(t)$.

i) $h(t) = 0$. Then $t = \lambda\vec{x}v^{\sigma}\vec{y}.v^{\sigma}$, where $\vec{x}$ and $\vec{y}$ are finite (possibly empty) lists of typed variables. First we regard the case where $\vec{x}$ is empty. If $\vec{y}$ is empty then the assignment to t is $I^{\sigma\to\sigma}$. Suppose that the assignment to $\lambda v^{\sigma}\vec{y}.v^{\sigma}$ is T, where $\vec{y} = y_1,\dots,y_n$. Then the assignment to $\lambda v^{\sigma}y^{\mu}\vec{y}.v^{\sigma}$ is $S(KK)T$ (we omit the subscripts). For $S(KK)T f^{\sigma} f^{\mu}\vec{f} = KK f^{\sigma}(T f^{\sigma})f^{\mu}\vec{f} = K(T f^{\sigma})f^{\mu}\vec{f} = T f^{\sigma}\vec{f} = f^{\sigma}$. Suppose the assignment to $\lambda\vec{x}v^{\sigma}\vec{y}.v^{\sigma}$ is U. Then KU is the assignment to $\lambda x^{\nu}\vec{x}v^{\sigma}\vec{y}.v^{\sigma}$, for $KUf^{\nu}\vec{f}f^{\sigma}\vec{g} = U\vec{f}f^{\sigma}\vec{g} = f^{\sigma}$.

ii) $h(t) > 0$. Now $t = \lambda\vec{x}t_1(t_2)$, where $\lambda\vec{x}t_1$, $\lambda\vec{x}t_2$ are closed terms. By the induction hypothesis the assignments to these terms are combinations T_1, T_2. Hence $T_i\vec{f}$ is the assignment to t_i, given that the assignment to $\vec{x}$ is $\vec{f}$, $i = 1,2$; and the assignment to $t_1(t_2)$ is $T_1\vec{f}(T_2\vec{f})$. It suffices to find a combination T such that $T\vec{f} = T_1\vec{f}(T_2\vec{f})$ for all $\vec{f}$. Such a T will be the assignment to t. We prove that T exists by induction on the length of $\vec{x}$. If $\vec{x}$ is empty we can let $T = T_1T_2$. Let $C(n)$ be the following claim: If p is the signature of a list $\vec{x}$ with length n of typed variables then there is a function P_p such that: i) $P_p(N_1,N_2)$ is a combination when N_1,N_2 are combinations of suitable types, ii) for all lists $\vec{f}$ with signature p : $P_p(N_1,N_2)\vec{f} = N_1\vec{f}(N_2\vec{f})$. $C(0)$ is true, for we can let $P_p(N_1,N_2) = N_1 N_2$, where p is the signature of the empty sequence. It is enough to prove that $C(n)$ is true for all n. For then we can let $T = P_p(T_1,T_2)$. Suppose $C(n)$ is true. To prove $C(n+1)$ let $\vec{x} = x_1,\ldots,x_n,x_{n+1}$ be a sequence with signature p. Now $N_1f_1,\ldots,f_{n+1}(N_2f_1 \ldots f_{n+1}) = S(N_1f_1 \ldots f_n)(N_2f_1 \ldots f_n)f_{n+1}$. Let $U_o = S$, $U_{k+1} = KU_k$. Then $U_nf_1,\ldots,f_n = S$, and by $C(n)$: $P_q(U_n,N_1)f_1,\ldots,f_n = S(N_1f_1 \ldots f_n)$, where q is the signature of $x_1,\ldots,x_n$. $P_q(P_q(U_n,N_1),N_2)f_1,\ldots,f_n = P_q(U_n,N_1)f_1,\ldots,f_n(N_2f_1 \ldots f_n)$ $= S(N_1f_1 \ldots f_n)(N_2f_1 \ldots f_n)$. Hence we can let $P_p(N_1,N_2) =$ $P_q(P_q(U_n,N_1),N_2)$. □

Translations between the recursion theories I, II, III, and IV.

Lemma 50: For all indexes e, all lists —— of objects from A : $\{e\}^{I}(\text{——}) \simeq \{e\}^{II}(\text{——})$.

Proof: By a straightforward induction on $|\{e\}^{I}(\text{——})|$ one can prove that if $\{e\}^{I}(\text{——}) \simeq t$ then $\{e\}^{II}(\text{——}) \simeq t$. Suppose $\{e\}^{II}(\text{——})\downarrow$. Since all objects in —— is of type 0 the schemes S15 and S16 has not been used in the computation tree of $\{e\}^{II}(\text{——})$. By induction on $|\{e\}^{II}(\text{——})|$ one easily proves that $\{e\}^{I}(\text{——}) \simeq \{e\}^{II}(\text{——})$. □

Lemma 51: For all indexes e, all lists —— of objects of pure type: $\{e\}^{II}(\text{——}) \simeq \{e\}^{III}(\text{——})$.

Proof: If $\{e\}^{II}(\text{——}) \simeq t$ then $\{e\}^{III}(\text{——}) \simeq t$ as in the proof of lemma 50. Suppose $\{e\}^{III}(\text{——})\downarrow$. Since each object in —— is of pure type, each subcomputation of $\{e\}^{III}(\text{——})$ has an argument list which consists of objects of pure type (proof by inspection of the schemes). By induction on $|\{e\}^{III}(\text{——})|$ one easily proves that $\{e\}^{II}(\text{——}) \simeq \{e\}^{III}(\text{——})$.

Suppose —— is the list $f_1,\ldots,f_n$, ——' is the list $f'_1,\ldots,f'_n$, where $f_i \in HC(\sigma_i)$, $f'_i \in HC(\sigma_i)$, $i = 1,\ldots,n$. We say that $\text{——} \leq \text{——}'$ if $f_i \leq f'_i$ for $i = 1,\ldots,n$.

<u>Lemma 52</u>: Suppose $\{e\}^{IV}(\text{——}) \simeq x$ and $\text{——} \leq \text{——}'$. Then $\{e\}^{IV}(\text{——}') \simeq x$, and $|\{e\}^{IV}(\text{——}')| \leq |\{e\}^{IV}(\text{——})|$.

<u>Corollary</u>: For each e $\lambda\text{——}\{e\}^{IV}(\text{——}) \in HC$.

Proof of lemma 52: The proof is by induction on $|\{e\}^{IV}(\text{——})|$. Induction hypothesis: $|\{e\}^{IV}(\text{——})| < \alpha$ and $\text{——} \leq \text{——}' \Rightarrow \{e\}^{IV}(\text{——}') \simeq \{e\}^{IV}(\text{——})$ and $|\{e\}^{IV}(\text{——}')| \leq |\{e\}^{IV}(\text{——})|$.

Suppose $|\{e\}^{IV}(\text{——})| = \alpha$, $\text{——} \leq \text{——}'$. If $\alpha = 0$ then it is easy to see that $\{e\}^{IV}(\text{——}') \simeq \{e\}^{IV}(\text{——})$ and $|\{e\}^{IV}(\text{——}')| = 0$. Suppose $\alpha > 0$. We prove the result only for the case S 16. Now $\{e\}^{IV}(f^\tau, \text{——}) \simeq f^\tau(p_1 \ldots p_k) \simeq t$, where $p_i = \lambda\vec{q}\{e_i\}^{IV}(\vec{q}, f^\tau, \text{——})$. If the type of p_i is not 0 let $(p_i)_\alpha$ be defined by: $(p_i)_\alpha(\vec{q}) \simeq \{e_i\}^{IV}(\vec{q}, f^\tau, \text{——})$ if $|\{e_i\}^{IV}(\vec{q}, f^\tau, \text{——})| < \alpha$, $(p_i)_\alpha(\vec{q}) \uparrow$ otherwise. If the type of p_i is 0 let $(p_i)_\alpha \simeq \{e_i\}^{IV}(f^\tau, \text{——})$ if $|\{e_i\}^{IV}(f^\tau, \text{——})| < \alpha$, $(p_i)_\alpha = \uparrow$ otherwise. As $|\{e\}^{IV}(f^\tau, \text{——})| = \alpha$ $(p_i)_\alpha \in HC$, $i = 1,\ldots,k$, and $f^\tau((p_1)_\alpha,\ldots,(p_k)_\alpha) \simeq t$. Suppose $f \leq f'$, $\text{——} \leq \text{——}'$. By the induction hypothesis $\{e_i\}^{IV}(\vec{q}, f', \text{——}') \simeq \{e_i\}^{IV}(\vec{q}, f, \text{——})$, and $|\{e_i\}^{IV}(\vec{q}, f', \text{——}')| \leq |\{e_i\}^{IV}(\vec{q}, f, \text{——})|$ when $(p_i)_\alpha(\vec{q}) \downarrow$. Let $p'_i = \lambda\vec{q}\{e_i\}^{IV}(\vec{q}, f', \text{——}')$. Let $(p'_i)_\alpha$ be defined in the same way as $(p_i)_\alpha$, $i = 1,\ldots,k$. By the induction hypothesis $(p'_i)_\alpha$ is a hereditarily consistent object, and $(p_i)_\alpha \leq (p'_i)_\alpha$. $f'((p'_i)_\alpha \ldots (p'_k)_\alpha) \simeq t$ because $f((p_1)_\alpha \ldots (p_k)_\alpha) \simeq t$. Hence $\{e\}^{IV}(f', \text{——}') \simeq t$, and $|\{e\}^{IV}(f', \text{——}')| \leq \alpha$. □

If —— is the list $\alpha_1,\dots,\alpha_k$, where $\alpha_i \in T(\sigma_i)$, $i = 1,\dots,k$, let $\Psi\,\text{——}$ denote the list $\Psi(\alpha_1),\dots,\Psi(\alpha_k)$.

<u>Lemma 53</u>: For all numbers e and all lists —— of total objects: $\{e\}^{III}(\text{——}) \simeq \{e\}^{IV}(\Psi\,\text{——})$, and $|\{e\}^{III}(\text{——})| = |\{e\}^{IV}(\Psi\,\text{——})|$ when $\{e\}^{III}(\text{——})\downarrow$.

Proof: Let —— be the list $\alpha_1,\dots,\alpha_k$. First we prove by induction on $|\{e\}^{III}(\text{——})|$ that if $\{e\}^{III}(\text{——}) \simeq t$ then $\{e\}^{IV}(\Psi\,\text{——}) \simeq t$ and $|\{e\}^{IV}(\Psi\,\text{——})| \leq |\{e\}^{III}(\text{——})|$. Then we prove by induction on $|\{e\}^{IV}(\Psi\,\text{——})|$ that if $\{e\}^{IV}(\Psi\,\text{——}) \simeq t$ then $\{e\}^{III}(\text{——}) \simeq t$ and $|\{e\}^{III}(\text{——})| \leq |\{e\}^{IV}(\Psi\,\text{——})|$.

The first induction. Induction hypothesis: $|\{e\}^{III}(\text{——})| < \nu \Rightarrow \{e\}^{III}(\text{——}) \simeq \{e\}^{IV}(\Psi\,\text{——})$, and $|\{e\}^{IV}(\Psi\,\text{——})| \leq |\{e\}^{III}(\text{——})|$. Suppose $|\{e\}^{III}(\text{——})| = \nu$. If $\nu = 0$ the result is trivial. So suppose $\nu > 0$. To prove the conclusion we regard several cases, one case for each scheme in III. We give the cases for S 9 and S 16 as examples.

S 9: $\{e\}^{III}(\text{——}) \simeq \{e_1\}^{III}(\{e_2\}^{III}(\text{——}),\text{——}) \simeq t$, and $|\{e_1\}^{III}(\{e_2\}^{III}(\text{——}),\text{——})| < \nu$, $|\{e_2\}^{III}(\text{——})| < \nu$. By the induction hypothesis $\{e_2\}^{IV}(\Psi\,\text{——}) \simeq \{e_2\}^{III}(\text{——})$, $|\{e_2\}^{IV}(\Psi\,\text{——})| < \nu$, $\{e_1\}^{IV}(\{e_2\}^{IV}(\Psi\,\text{——}),\Psi\,\text{——}) \simeq t$, $|\{e_1\}^{IV}(\{e_2\}^{IV}(\Psi\,\text{——}),\Psi\,\text{——})| < \nu$. Hence $\{e\}^{IV}(\Psi\,\text{——}) \simeq t$ and $|\{e\}^{IV}(\Psi\,\text{——})| \leq \nu$.

S 16: $\{e\}^{III}(\alpha^\tau,\text{——}) \simeq \alpha^\tau(\delta_1 \dots \delta_k) \simeq t$, where $\delta_i = \lambda\vec{\gamma}\{e_i\}^{III}(\vec{\gamma},\alpha^\tau,\text{——})$. As $\alpha^\tau(\delta_1 \dots \delta_k)\downarrow$ $\delta_1,\dots,\delta_k$ are total objects. $|\{e_i\}^{III}(\vec{\gamma},\alpha^\tau,\text{——})| < \nu$ for all $\vec{\gamma}$ because $|\{e\}^{III}(\alpha^\tau,\text{——})| = \nu$. By the induction hypothesis $\{e_i\}^{IV}(\Psi\vec{\gamma},\Psi\alpha^\tau,\Psi\,\text{——}) \simeq \{e_i\}^{III}(\vec{\gamma},\alpha^\tau,\text{——})$, and $|\{e_i\}^{IV}(\Psi\vec{\gamma},\Psi\alpha^\tau,\Psi\,\text{——})| < \nu$. $\{e\}^{IV}(\Psi\alpha,\Psi\,\text{——}) \simeq \Psi\alpha(p_1 \dots p_k) \simeq \alpha(\Phi p_1 \dots \Phi p_k)$, where $p_i = \lambda\vec{q}\{e_i\}^{IV}(\vec{q},\Psi\alpha,\Psi\,\text{——})$. By lemma 52 $p_i \in HC$, $i = 1,\dots,k$. If the type of p_i is not 0 then $\Phi(p_i)(\vec{\gamma}) \simeq \{e_i\}^{IV}(\Psi\vec{\gamma},\Psi\alpha,\Psi\,\text{——}) \simeq \{e_i\}^{III}(\vec{\gamma},\alpha,\text{——})$ (the last equality by the induction hypothesis).

Hence $\Phi p_i = \delta_i$, and $\{e\}^{IV}(\Psi\alpha,\Psi \text{---}) \simeq \alpha(\delta_1 \dots \delta_k) \simeq t$. $|\{e\}^{IV}(\Psi\alpha,\Psi \text{---})|$ is the least ordinal μ such that $(p_i)_\mu \in HC$, $i = 1,\dots,k$, and $\Psi(\alpha)((p_1)_\mu \dots (p_k)_\mu)\downarrow$. Hence $\mu \leq \nu$, i.e. $|\{e\}^{IV}(\Psi\alpha,\Psi \text{---})| \leq \nu$.

The second induction. Induction hypothesis: $|\{e\}^{IV}(\Psi \text{---})| < \nu \Rightarrow \{e\}^{III}(\text{---}) \simeq \{e\}^{IV}(\Psi \text{---})$ and $|\{e\}^{III}(\text{---})| \leq |\{e\}^{IV}(\Psi \text{---})|$. Suppose $|\{e\}^{IV}(\Psi \text{---})| = \nu$. We prove the result only for case S 16. Now $|\{e\}^{IV}(\Psi\alpha,\Psi \text{---})| = \nu$, $\{e\}^{IV}(\Psi\alpha,\Psi \text{---}) \simeq \Psi(\alpha)(p_1 \dots p_k) \simeq t$, where $p_i = \lambda\vec{q}\{e_i\}^{IV}(\vec{q},\Psi\alpha,\Psi \text{---})$. $\Psi\alpha((p_1)_\nu \dots (p_k)_\nu) \simeq t$ because $|\{e\}^{IV}(\Psi\alpha,\Psi \text{---})| = \nu$. Hence $\alpha(\Phi(p_1)_\nu \dots \Phi(p_k)_\nu) \simeq t$, and $\Phi(p_i)_\nu$ is a total object. So $\Phi(p_i)_\nu(\vec{\gamma})\downarrow$ for all $\vec{\gamma}$, i.e. $(p_i)_\nu(\Psi\vec{\gamma})\downarrow$, $|\{e_i\}^{IV}(\Psi\vec{\gamma},\Psi\alpha,\Psi \text{---})| < \nu$. By the induction hypothesis $\{e_i\}^{III}(\vec{\gamma},\alpha,\text{---}) \simeq \{e_i\}^{IV}(\Psi\vec{\gamma},\Psi\alpha,\Psi \text{---})$, and $|\{e_i\}^{III}(\vec{\gamma},\alpha,\text{---})| < \nu$ for all $\vec{\gamma}$. Let $\delta_i = \lambda\vec{\gamma}\{e_i\}^{III}(\vec{\gamma},\alpha,\text{---})$. Now $\Phi(p_i)_\nu = \delta_i$, $\alpha(\delta_1 \dots \delta_k) \simeq t$. It follows that $\{e\}^{III}(\alpha,\text{---}) \simeq t$, and $|\{e\}^{III}(\alpha,\text{---})| \leq \nu$. $\square$

<u>To imbed $\mathcal{A}_\omega(\mathcal{B})$ into IV.</u>

<u>Lemma 54</u>: There is a primitive recursive function δ_1 such that for all lists ---, === of hereditarily consistent objects, all e: $\{e\}^{IV}(\text{---}) \simeq \{\delta_1(e,m)\}^{IV}(\text{---},\text{===})$, and if $\{e\}^{IV}(\text{---})\downarrow$ then $|\{\delta_1(e,m)\}^{IV}(\text{---},\text{===})| \geq |\{e\}^{IV}(\text{---})|$. ($m$ is the signature of the list ===.)

Proof: This is an application of the recursion theorem for primitive recursive functions. We define a primitive recursive function $t(e,m,s)$ by cases. There is one case for each scheme in IV, and one otherwise case. We give the cases S 9 and S 12 as examples.

S 9: $e = \langle 9,n,e_1,e_2\rangle$. Let $t(e,m,s) = \langle 9,n',t(e_1,m,s),t(e_2,m,s)\rangle$ where n' is the signature of the list $\text{---},\text{===}$.

S 12: $e = \langle 12,n\rangle$. $\{e\}^{IV}(x,\text{---}) \simeq \{x\}^{IV}(\text{---})$. Choose $t(e,m,s)$

such that $\{t(e,m,s)\}^{IV}(x, —, ═) \simeq \{\{s\}_{PR}(x,m)\}^{IV}(—, ═)$, where $\{s\}_{PR}$ denotes the primitive recursive function with index s.

By the recursion theorem for primitive recursive functions there is an s such that $t(e,m,s) = \{s\}_{PR}(e,m)$ for all e,m. Let $\delta_1(e,m) = t(e,m,s)$ for this s. By induction on $|\{e\}^{IV}(—)|$ one can prove that if $\{e\}^{IV}(—) \simeq t$ then $\{\delta_1(e,m)\}^{IV}(—,═) \simeq t$, and $|\{e\}^{IV}(—)| \leq |\{\delta_1(e,m)\}^{IV}(—, ═)|$. By induction on $|\{\delta_1(e,m)\}^{IV}(—, ═)|$ one can prove that if $\{\delta_1(e,m)\}^{IV}(—, ═) \simeq t$ then $\{e\}^{IV}(—) \simeq t$.

□

<u>Theorem 23</u>: There is a primitive recursive function δ_2 such that for all e_1, e_2, all lists $—$, $═$ of hereditarily consistent objects: $\{\delta_2(e_1,e_2,j)\}^{IV}(—, ═) \simeq \{e_1\}^{IV}(—, f^{\tau}, ═)$, where $f^{\tau} = \lambda\vec{g}\{e_2\}^{IV}(\vec{g}, —, ═)$, and f^{τ} is the j-th element in the argument list of $\{e_1\}^{IV}$.

Proof: The proof is by induction on the level of τ. For $l \geq 0$ we define a primitive recursive function $\delta(e_1,e_2,j,l)$ such that for all $e_1, e_2, —, ═$: $\{\delta(e_1,e_2,j,l)\}^{IV}(—, ═) \simeq \{e_1\}^{IV}(—, f^{\tau}, ═)$ if $l(\tau) = l$. Finally we let $\delta_2(e_1,e_2,j) = \delta(e_1,e_2,j,l)$, where the number l can be obtained from e_1 and j.

If $\tau = 0$, $j = 1$ then the substitution is just S 9. Let $\delta(e_1,e_2,1,0) = \langle 9,n,e_1,e_2\rangle$, where n is the signature of the list $═$ (the list $—$ is empty because $j = 1$). If $j > 1$ we move the j-th element in the argument list of $\{e_1\}^{IV}$ to the front of the list (apply S 11), and then substitute. Let $\delta(e_1,e_2,j,0) = \delta(i,e_2,1,0)$, where $\{i\}^{IV}(x, —, ═) \simeq \{e_1\}^{IV}(—, x, ═)$.

Suppose we have defined $\delta(e_1,e_2,j,l)$ for all e_1,e_2,j. To define $\delta(e_1,e_2,j,l+1)$ suppose $l(\tau) = l+1$. We define a primitive recursive function $t(e_1,e_2,j,s)$. By the recursion theorem for primitive recursive functions there is an s such that $t(e_1,e_2,j,s) = \{s\}_{PR}(e_1,e_2,j)$ for all e_1,e_2,j. We let $\delta(e_1,e_2,j,l+1) = t(e_1,e_2,j,s)$ for this s. The definition of t is by cases. There is one case for each scheme in IV, and one otherwise case. We give the cases S1, S9, S11, S12 and S16 as examples. We assume that the j-th object in the argument list of $\{e_1\}^{\mathrm{IV}}$ is of a type with level $l+1$. Otherwise let $t(e_1,e_2,j,s) = 0$.

$$S1: \quad \{e_1\}^{\mathrm{IV}}(x, \text{---}, f^{\tau}, \text{===}) \simeq \begin{cases} 0 & \text{if } x \in N \\ 1 & \text{if } x \notin N \end{cases}$$

Let $t(e_1,e_2,j,s) = \langle 1,n' \rangle$, where n' is the signature of the list $x, \text{---}, \text{===}$.

$$S9: \quad \{e_1\}^{\mathrm{IV}}(\text{---}, f^{\tau}, \text{===}) \simeq \{i_1\}^{\mathrm{IV}}(\{i_2\}^{\mathrm{IV}}(\text{---}, f^{\tau}, \text{===}), \text{---}, f^{\tau}, \ ).$$

Let $t(e_1,e_2,j,s) = \langle 9,n',t(i_1,e_2,j+1,s),t(i_2,e_2,j,s) \rangle$, where n' is the signature of $\text{---}, \text{===}$.

S11: $\{e_1\}^{\mathrm{IV}}(\text{---}, f^{\tau}, \text{===}) \simeq \{i\}^{\mathrm{IV}}(\text{---}')$, where $\text{---}'$ is obtained from $\text{---}, f^{\tau}, \text{===}$ by moving the m-th element in $\text{---}, f^{\tau}, \text{===}$ to the front of the list. If $m < j$ let $t(e_1,e_2,j,s) = t(i,e_2,j,s)$. If $m = j$ let $t(e_1,e_2,j,s) = t(i,e_2,1,s)$. If $m > j$ let $t(e_1,e_2,j,s) = t(i,e_2,j+1,s)$.

$$S12: \quad \{e_1\}^{\mathrm{IV}}(x, \text{---}, f^{\tau}, \text{===}) \simeq \{x\}^{\mathrm{IV}}(\text{---}, f^{\tau}, \text{===}).$$

Choose $t(e_1,e_2,j,s)$ such that $\{t(e_1,e_2,j,s)\}^{\mathrm{IV}}(x, \text{---}, \text{===}) \simeq \{\{s\}_{PR}(x,e_2,j-1)\}^{\mathrm{IV}}(\text{---}, \text{===})$.

S16: $\{e_1\}^{\mathrm{IV}}(f^{\nu}, \text{---}) \simeq f^{\nu}(p_1 \dots p_k)$, where $p_i = \lambda\vec{q}\{e_i'\}^{\mathrm{IV}}(\vec{q}, f^{\nu}, \text{---})$. First we regard the case $j = 1$. Then $\nu = \tau$, and we substitute $\lambda\vec{g}\{e_2\}^{\mathrm{IV}}(\vec{g}, \text{---})$ for f^{τ} in $\{e_1\}^{\mathrm{IV}}(f^{\tau}, \text{---})$. First we substitute for f^{τ} in $\{e_i'\}^{\mathrm{IV}}(\vec{q}, f^{\tau}, \text{---})$. To do that we find an index e_{2i} such that $\{e_{2i}\}^{\mathrm{IV}}(\vec{g},\vec{q}, \text{---}) \simeq \{e_2\}^{\mathrm{IV}}(\vec{g}, \text{---})$ for all $\vec{g},\vec{q}, \text{---}$ (lemma 54 and applications of S11). Let $e_i'' = t(e_i',e_{2i},j_i,s)$, where $j_i - 1 =$ the length of the sequence $\vec{q}$ in $\{e_i'\}^{\mathrm{IV}}(\vec{q}, f^{\tau}, \text{---})$. When t is defined and we have chosen the correct value of s we will have: $\{e_i''\}^{\mathrm{IV}}(\vec{q}, \text{---}) \simeq \{e_i'\}^{\mathrm{IV}}(\vec{q}, f^{\tau}, \text{---})$ for $i = 1,\dots,k$. We now substitute $\lambda\vec{q}\{e_i''\}^{\mathrm{IV}}(\vec{q}, \text{---})$ for g_i in $\{e_2\}^{\mathrm{IV}}(g_1 \dots g_k, \text{---})$. As the level of the type of g_i is not greater than l indeces for these substitutions can be found by $\delta(e_1,e_2,j,l)$. From this we can construct $t(e_1,e_2,1,s)$.

If $j > 1$ we substitute $\lambda\vec{g}\{e_2\}^{IV}(\vec{g},f^{\nu},\text{---},\text{===})$ for f^{τ} in $\{e_1\}^{IV}(f^{\nu},\text{---},f^{\tau},\text{===})$. First substitute for f^{τ} in $\{e_i'\}^{IV}(\vec{q},f^{\nu},\text{---},f^{\tau},\text{===})$. To do that we find an index e_{2i} such that $\{e_{2i}\}^{IV}(\vec{g},\vec{q},f^{\nu},\text{---},\text{===}) \simeq \{e_2\}^{IV}(\vec{g},f^{\nu},\text{---},\text{===})$. Let $e_i'' = t(e_i',e_{2i},j_i,s)$, where $j_i - 1$ is the length of the sequence $\vec{q},f^{\nu},\text{---}$ in $\{e_i'\}^{IV}(\vec{q},f^{\nu},\text{---},f^{\tau},\text{===})$. Let $t(e_1,e_2,j,s) = h(\langle 16,n,\langle e_1''\ldots e_k''\rangle\rangle)$, where h is a primitive recursive function such that for all e, ---: $\{e\}^{IV}(\text{---}) \simeq \{h(e)\}^{IV}(\text{---})$, $|\{e\}^{IV}(\text{---})| = \lambda \Rightarrow |\{h(e)\}^{IV}(\text{---})| \geq \lambda + 1$.

Choose s such that $t(e_1,e_2,j,s) = \{s\}_{PR}(e_1,e_2,j)$ for all e_1,e_2,j, and let $\delta(e_1,e_2,j,l+1) = t(e_1,e_2,j,s)$ for this s. It remains to prove that $\{\delta(e_1,e_2,j,l+1)\}^{IV}(\text{---},\text{===}) \simeq \{e_1\}^{IV}(\text{---},f^{\tau},\text{===})$, where $f^{\tau} = \lambda\vec{g}\{e_2\}^{IV}(\vec{g},\text{---},\text{===})$. In fact we will prove the stronger result $C(l+1)$, where $C(l+1)$ is the statement:

i) $\{e_1\}^{IV}(\text{---},f^{\tau},\text{===}) \simeq t \Rightarrow \{\delta(e_1,e_2,j,l+1)\}^{IV}(\text{---},\text{===}) \simeq t$,

ii) $\{\delta(e_1,e_2,j,l+1)\}^{IV}(\text{---},\text{===}) \simeq t \Rightarrow \{e_1\}^{IV}(\text{---},f^{\tau},\text{===})\downarrow$,

and there is an object $p \in HC(\tau)$ such that $p \leq f^{\tau}$, $\{e_1\}^{IV}(\text{---},p,\text{===})\downarrow$, $|\{\delta(e_1,e_2,j,l+1)\}^{IV}(\text{---},\text{===})| > \sup\{|\{e_2\}^{IV}(\vec{g},\text{---},\text{===})| : p(\vec{g})\downarrow\}$, $|\{\delta(e_1,e_2,j,l+1)\}^{IV}(\text{---},\text{===})| > |\{e_1\}^{IV}(\text{---},p,\text{===})|$.

The corresponding result for $l = o$, denoted by $C(0)$, can easily by proved:

i) $\{e_1\}^{IV}(\text{---},f^{o},\text{===}) \simeq t \Rightarrow \{\delta(e_1,e_2,j,0)\}^{IV}(\text{---},\text{===}) \simeq t$,

ii) $\{\delta(e_1,e_2,j,0)\}^{IV}(\text{---},\text{===}) \simeq t \Rightarrow \{e_1\}^{IV}(\text{---},f^{o},\text{===})\downarrow$,

and $|\{\delta(e_1,e_2,j,0)\}^{IV}(\text{---},\text{===})| > |\{e_1\}^{IV}(\text{---},f^{o},\text{===})|$, $|\{\delta(e_1,e_2,j,0)\}^{IV}(\text{---},\text{===})| > |\{e_2\}^{IV}(\text{---},\text{===})|$.

First induction hypothesis: $C(l')$ is true when $l' \leq l$. Second induction hypothesis: $|\{e_1\}^{IV}(\text{---},f^{\tau},\text{===})| < \lambda \Rightarrow \{\delta(e_1,e_2,j,l+1)\}^{IV}(\text{---},\text{===}) \simeq \{e_1\}^{IV}(\text{---},f^{\tau},\text{===})$.

Suppose $\{e_1\}^{IV}(\text{—},f^\tau,\text{==}) \simeq t$, $|\{e_1\}^{IV}(\text{—},f^\tau,\text{==})| = \lambda$. We prove that $\{\delta(e_1,e_2,j,l+1)\}^{IV}(\text{—},\text{==}) \simeq t$ only for the case S 16.

Now $\{e_1\}^{IV}(f^\nu,\text{—}) \simeq f^\nu(p_1 \ldots p_k) \simeq t$, where $p_i = \lambda\vec{q}\{e_i'\}^{IV}(\vec{q},f^\nu,\text{—})$. First we regard the case where $j = 1$, i.e. $\nu = \tau$ and we substitute $\lambda\vec{g}\{e_2\}^{IV}(\vec{g},\text{—})$ for f^ν. As $|\{e_1\}^{IV}(f^\nu,\text{—})| = \lambda$ we have that $f((p_1)_\lambda \ldots (p_k)_\lambda) \simeq t$, where $(p_i)_\lambda$ is defined by: If the type of p_i is not 0 let $(p_i)_\lambda(\vec{q}) \simeq \{e_i'\}^{IV}(\vec{q},f^\nu,\text{—})$ if $|\{e_i'\}^{IV}(\vec{q},f^\nu,\text{—})| < \lambda$, $(p_i)_\lambda(\vec{q})\uparrow$ otherwise. If the type of p_i is 0 let $(p_i)_\lambda \simeq \{e_i'\}^{IV}(f^\nu,\text{—})$ if $|\{e_i'\}^{IV}(f^\nu,\text{—})| < \lambda$, $(p_i)_\lambda = \uparrow$ otherwise. By the second induction hypothesis $\{e_i''\}^{IV}(\vec{q},\text{—}) \simeq \{e_i'\}^{IV}(\vec{q},f^\nu,\text{—})$ when $(p_i)_\lambda(\vec{q})\downarrow$. Hence $(p_i)_\lambda \leq \lambda\vec{q}\{e_i''\}^{IV}(\vec{q},\text{—})$, and

$f^\nu(\lambda\vec{q}\{e_1''\}^{IV}(\vec{q},\text{—}),\ldots,\lambda\vec{q}\{e_k''\}^{IV}(\vec{q},\text{—})) \simeq t$,

$\{e_2\}^{IV}(\lambda\vec{q}\,\{e_1''\}^{IV}(\vec{q},\text{—}),\ldots,\lambda\vec{q}\{e_k''\}^{IV}(\vec{q},\text{—}),\text{—}) \simeq t$.

The level of the type of $\lambda\vec{q}\{e_i''\}^{IV}(\vec{q},\text{—})$ is not greater than l. By the first induction hypothesis $\{\delta(e_1,e_2,1,l+1)\}^{IV}(\text{—}) \simeq t$.

The case $j > 1$: Now $\{e_1\}^{IV}(f^\nu,\text{—},f^\tau,\text{==}) \simeq f^\nu(p_1 \ldots p_k) \simeq t$, where $p_i = \lambda\vec{q}\{e_i'\}^{IV}(\vec{q},f^\nu,\text{—},f^\tau,\text{==})$.
Let $(p_i)_\lambda$ be defined as above. Then $f^\nu((p_1)_\lambda \ldots (p_k)_\lambda) \simeq t$. By the second induction hypothesis $\{e_i''\}^{IV}(\vec{q},f^\nu,\text{—},\text{==}) \simeq \{e_i'\}^{IV}(\vec{q},f^\nu,\text{—},f^\tau,\text{==})$ when $(p_i)_\lambda(\vec{q})\downarrow$. Hence $(p_i)_\lambda \leq \lambda\vec{q}\{e_i''\}^{IV}(\vec{q},f^\nu,\text{—},\text{==})$,

$f^\nu(\lambda\vec{q}\{e_1''\}^{IV}(\vec{q},f^\nu,\text{—},\text{==}),\ldots,\lambda\vec{q}\{e_k''\}^{IV}(\vec{q},f^\nu,\text{—},\text{==})) \simeq t$,

$\{\delta(e_1,e_2,1,l+1)\}^{IV}(f^\nu,\text{—},\text{==}) \simeq t$.

Now we have proved i) in the claim $C(l+1)$. To prove ii) we assume as our third induction hypothesis: $|\{\delta(e_1,e_2,j,l+1)\}^{IV}(\text{—},\text{==})| < \lambda$ $\Longrightarrow$ $\{e_1\}^{IV}(\text{—},f^\tau,\text{==}) \simeq \{\delta(e_1,e_2,j,l+1)\}^{IV}(\text{—},\text{==})$, and there is an object $p \in HC(\tau)$ such that $p \leq f^\tau$, $\{e_1\}^{IV}(\text{—},p,\text{==})\downarrow$,

$|\{\delta(e_1,e_2,j,l+1)\}^{IV}(\text{—},\text{==})| > \sup\{|\{e_2\}^{IV}(\vec{g},\text{—},\text{==})| : p(\vec{g})\downarrow\}$,

$|\{\delta(e_1,e_2,j,l+1)\}^{IV}(\text{—},\text{==})| > |\{e_1\}^{IV}(\text{—},p,\text{==})|$.

Assume $\{\delta(e_1,e_2,j,l+1)\}^{IV}(\text{—},\text{==}) \simeq t$, $|\{\delta(e_1,e_2,j,l+1)\}^{IV}(\text{—},\text{==})| = \lambda$. We prove the conclusion only for the case S 16. First we regard the case $j = 1$. Now $\{e_1\}^{IV}(f^\tau, —) \simeq f^\tau(p_1 \dots p_k)$, and we substitute $\lambda\vec{g}\{e_2\}^{IV}(\vec{g}, —)$ for f^τ. Now $\{\delta(e_1,e_2,1,l+1)\}^{IV}(—) \simeq \{e_2\}^{IV}(\lambda\vec{q}\,\{e_1''\}^{IV}(\vec{q}, —),\dots,\lambda\vec{q}\{e_k''\}(\vec{q},—),—) \simeq t$ by the first induction hypothesis, and there are subfunctions g_i' of $g_i = \lambda\vec{q}\{e_i''\}^{IV}(\vec{q}, —)$ such that $\{e_2\}^{IV}(g_1' \dots g_k', —) \simeq t$, $|\{\delta(e_1,e_2,1,l+1)\}^{IV}(—)| > |\{e_2\}^{IV}(g_1' \dots g_k', —)|$, $|\{\delta(e_1,e_2,1,l+1)\}^{IV}(—)| > \sup\{|\{e_i''\}^{IV}(\vec{q}, —)| : g_i'(\vec{q})\downarrow\}$ for $i = 1 \dots k$. Let μ be an ordinal such that $\mu < \lambda$ and $\mu \geq \sup\{|\{e_i''\}^{IV}(\vec{q}, —)| : g_i'(\vec{q})\downarrow\}$, $i = 1 \dots k$. Suppose $g_i'(\vec{q})\downarrow$. It follows from the third induction hypothesis that there is an object $p_i(\vec{q}) \in HC(\tau)$ such that $p_i(\vec{q}) \leq f^\tau$, $\{e_i'\}^{IV}(\vec{q}, p_i(\vec{q}), —) \simeq \{e_i''\}^{IV}(\vec{q},—)$, $\mu > |\{e_i'\}^{IV}(\vec{q}, p_i(\vec{q}), —)|$, $\mu > \sup\{|\{e_2\}^{IV}(\vec{g}, —)| : p_i(\vec{q})(\vec{g})\downarrow\}$.

It remains to find an object p in $HC(\tau)$ such that $p \leq f^\tau$, $\{e_1\}^{IV}(p, —) \simeq t$, $|\{e_1\}^{IV}(p, —)| < \lambda$, and $\lambda > \sup\{|\{e_2\}^{IV}(\vec{g}, —)| : p(\vec{g})\downarrow\}$. Define p as follows: Let $p_i = \cup\,\{p_i(\vec{q}) : g_i'(\vec{q})\downarrow\}$. Let $p' = p_1 \cup p_2 \cup \dots \cup p_k$. Let p be the least hereditarily consistent object such that $p' \leq p$ and $p(g_1' \dots g_k') \simeq t$. There is no conflict in this definition as $p_i(\vec{q}) \leq f^\tau$ for all $i,\vec{q}$, and $f^\tau(g_1' \dots g_k') \simeq t$. So $p \leq f^\tau$. To prove that $\lambda > \sup\{|\{e_2\}^{IV}(\vec{g}, —)| : p(\vec{g})\downarrow\}$ suppose $p(\vec{g})\downarrow$. If $g_1',\dots,g_k' \leq \vec{g}$ then $|\{e_2\}^{IV}(\vec{g}, —)| < \lambda$ because $|\{e_2\}^{IV}(g_1' \dots g_k', —)| < \lambda$. If $p(\vec{g}) \simeq p_i(\vec{q})(\vec{p})$, where $g_i'(\vec{q})\downarrow$, then $\mu > |\{e_2\}^{IV}(\vec{g}, —)|$ as mentioned above. To prove that $\{e_1\}^{IV}(p, —) \simeq t$ and $|\{e_1\}^{IV}(p, —)| < \lambda$ let $(r_i)_\mu$ be defined by: $(r_i)_\mu(\vec{q}) \simeq \{e_i'\}^{IV}(\vec{q},p, —)$ if $|\{e_i'\}^{IV}(\vec{q},p, —)| < \mu$, $(r_i)_\mu(\vec{q})\uparrow$ otherwise. By the results above $g_i' \leq (r_i)_\mu$. Hence $\{e_2\}^{IV}((r_1)_\mu,\dots,(r_k)_\mu, —) \simeq t$ because $\{c_2\}^{IV}(g_1',\dots,g_k', —) \simeq t$. Hence $\{e\}^{IV}(p, —) \simeq t$, and $|\{e\}^{IV}(p, —)| \leq \mu$.

Finally we regard the case $j > 1$. Suppose $\{\delta(e_1,e_2,j,l+1)\}^{IV}(f^\nu, —, ═) \simeq t$, $|\{\delta(e_1,e_2,j,l+1)\}^{IV}(f^\nu, —, ═)| = \lambda$. By the construction of $t(e_1,e_2,j,s)$ there is an ordinal μ such that $\mu < \lambda$, $|\{\langle 16,n,\langle e_1'' \dots e_k''\rangle\rangle\}^{IV}(f^\nu, —, ═)| = \mu$,

$\{\langle 16,n,\langle c_1'' \ldots e_k''\rangle\rangle\}^{IV}(f^\nu, —, ==) \simeq t$. Hence $f^\nu(\lambda\vec{q}\{e_1''\}^{IV}(\vec{q},f^\nu,—,==)$, $\ldots, \lambda\vec{q}\{e_k''\}^{IV}(\vec{q},f^\nu,—,==)) \simeq t$, and $f^\nu((r_1)_\mu \ldots (r_k)_\mu) \simeq t$, where $(r_i)_\mu(\vec{q}) \simeq \{e_i''\}^{IV}(\vec{q},f^\nu,—,==)$ if $|\{e_i''\}^{IV}(\vec{q},f^\nu,—,==)| < \mu$, $(r_i)_\mu(\vec{q})\uparrow$ otherwise. Suppose $(r_i)_\mu(\vec{q})\downarrow$. By the third induction there is a subfunction $p_i(\vec{q})$ of f^τ such that
$\{e_i''\}^{IV}(\vec{q},f^\nu,—,==) \simeq \{e_i'\}^{IV}(\vec{q},f^\nu,—,p_i(\vec{q}),==)$,
$|\{e_i''\}^{IV}(\vec{q},f^\nu,—,==)| > |\{e_i'\}^{IV}(\vec{q},f^\nu,—,p_i(\vec{q}),==)|$,
$|\{e_i''\}^{IV}(\vec{q},f^\nu,—,==)| > \sup\{|\{e_2\}^{IV}(\vec{g},f^\nu,—,==)| : p_i(\vec{q})(\vec{g})\downarrow\}$. Hence $\mu > |\{e_i'\}^{IV}(\vec{q},f^\nu,—,p_i(\vec{q}),==)|$, $\mu > \sup\{|\{e_2\}^{IV}(\vec{g},f^\nu,—,==)| :$ $p_i(\vec{q})(\vec{g})\downarrow\}$.

Let $p \in HC(\tau)$ be defined by: $p = p_1 \cup p_2 \cup \ldots \cup p_k$, where $p_i =$ $\cup\{p_i(\vec{q}) : (r_i)_\mu(q)\downarrow\}$. It remains to prove that $\{e_1\}^{IV}(f^\nu,—,p,==)$ $\simeq t$, $\lambda > \sup\{|\{e_2\}^{IV}(\vec{g},f^\nu,—,==)| : p(\vec{g})\downarrow\}$, $\lambda > \{e_1\}^{IV}(f^\nu,—,p,==)|$. Suppose $p(\vec{g})\downarrow$. Then $p_i(\vec{q})(\vec{q})\downarrow$ for some $i,\vec{q}$ such that $(r_i)_\mu(q)\downarrow$. By the results above $\mu > |\{e_2\}^{IV}(\vec{g},f^\nu,—,==)|$, hence $\mu \geq$ $\sup\{|\{e_2\}^{IV}(\vec{g},f^\nu,—,==)| : p(\vec{g})\downarrow\}$. To prove that $\{e_1\}^{IV}(f^\nu,—,p,==)$ $\simeq t$, $|\{e_1\}^{IV}(f^\nu,—,p,==)| < \lambda$ let $r_i' = \lambda\vec{q}\{e_i'\}^{IV}(\vec{q},f^\nu,—,p,==)$. By the results above $(r_i)_\mu \leq (r_i')_\mu$. Hence $f^\nu((r_1')_\mu \ldots (r_k')_\mu) \simeq t$ as $f((r_1)_\mu,\ldots,(r_k)_\mu) \simeq t$. Hence $\{e\}^{IV}(f^\nu,—,p,==) \simeq t$, and $|\{e\}^{IV}(f^\nu,—,p,==)| \leq \mu$. □

<u>Lemma 55</u>: For each type symbol τ there is an index e such that for all $h \in HC(\tau)$: $h = \lambda\vec{g}\{e\}^{IV}(h,\vec{g})$.

Proof: The proof is by induction on the level of τ. If $\tau = 0$ let e be the index of S 2 with the empty list for —. Then $\{e\}^{IV}(x) = x$. Suppose that the lemma is true when $l(\tau) \leq l$. Suppose $l(\tau) = l+1$. Let $\tau_1,\ldots,\tau_k$ be the factors of τ. Then $l(\tau_i) \leq l$, $i = 1,\ldots,k$. By the induction hypothesis there are indeces $e_1,\ldots,e_k$ such that $\lambda\vec{p}\{e_i\}^{IV}(g^{\tau_i},\vec{p}) = g^{\tau_i}$ for all $g^{\tau_i} \in HC(\tau_i)$. By lemma 54 and applications of S 11 there are indeces $e_1',\ldots,e_k'$ such that $\{e_i'\}^{IV}(\vec{p},h,g^{\tau_1}\ldots g^{\tau_k})$

$\simeq \{e_i\}^{IV}(g^{\tau_i},\vec{p})$ for all $\vec{p}, h, g^{\tau_1},\ldots,g^{\tau_k}$.

Let $e = \langle 16, n, \langle e_1' \ldots e_k' \rangle\rangle$, where n is the signature of the list $\tau\,\tau_1\,\tau_2,\ldots,\tau_k$. Then $\{e\}^{IV}(h, g^{\tau_1} \ldots g^{\tau_k}) \simeq h(\lambda\vec{p}\{e_1'\}^{IV}(\vec{p},h,\vec{g}),\ldots,$ $\lambda\vec{p}\{e_k'\}^{IV}(\vec{p},h,\vec{g})) \simeq h(\vec{g})$. □

Lemma 56: For each natural number n there is a primitive recursive function S^n such that for all natural numbers $e, x_1,\ldots,x_n$, all lists —— : $\{S^n(e,x_1 \ldots x_n)\}^{IV}(\text{——}) \simeq \{e\}^{IV}(x_1 \ldots x_n, \text{——})$, and $|\{S^n(e,x_1 \ldots x_n)\}^{IV}(\text{——})| > |\{e\}^{IV}(x_1 \ldots x_n, \text{——})|$ when $\{e\}^{IV}(x_1 \ldots x_n, \text{——})\downarrow$.

Proof: Let $S^0(e) = e$, $S^{n+1}(e,x_1 \ldots x_{n+1}) = \langle 9, m, S^n(e,x_1 \ldots x_n),$ $\delta(x_{n+1},m)\rangle$, where m is the signature of the list —— (m can be obtained from e and n), and $\delta(x,m)$ is a primitive recursive function such that $\{\delta(x,m)\}^{IV}(x, \text{——}) = x$ if $x \in N$, $= 0$ if $x \notin N$, for all lists —— with signature m. □

Theorem 24: Suppose $h_1,\ldots,h_k \in HC$, and $f \in \mathcal{R}_\omega(\{h_1 \ldots h_k\})$. Then there is an index e such that $f = \lambda\vec{g}\{e\}^{IV}(\vec{g},h_1 \ldots h_k)$.

Corollary: Suppose $\mathcal{B} \subseteq HC$, $f \in \mathcal{R}_\omega(\mathcal{B})$. Then there is an index e and objects $h_1,\ldots,h_k \in \mathcal{B}$ such that $f = \lambda\vec{g}\{e\}^{IV}(\vec{g},h_1 \ldots h_k)$.

Proof of theorem 24: First we find indeces for each initial function $h_1,\ldots,h_k$, I, K, S, DC, FP. Then we prove that there is an index for $f_1^{\sigma\to\tau}(f_2^\sigma)$ given indeces for f_1 and f_2.

i. Index for h_i. By lemma 55 there is an index e_i such that $h_i = \lambda\vec{g}\{e_i\}^{IV}(h_i,\vec{g})$.

ii. Index for $I^{\sigma\to\sigma}$. By lemma 55 there is an index e such that $\{e\}^{IV}(f^\sigma,\vec{g}) \simeq f^\sigma(\vec{g})$. Hence $I^{\sigma\to\sigma} = \lambda f^\sigma\vec{g}\{e\}^{IV}(f^\sigma,\vec{g})$.

iii. Index for $K^{\sigma\to\tau\to\sigma}$. $K(f^\sigma,g^\tau) = f^\sigma$. By lemma 55 there is an index e such that $\{e\}^{IV}(f^\sigma,\vec{h}) \simeq f^\sigma(\vec{h})$. Using lemma 54 and the scheme S 11 one can find an index c' such that $\{e'\}^{IV}(f^\sigma,g^\tau,\vec{h}) = \{e\}^{IV}(f^\sigma,\vec{h})$ for all $f, g, \vec{h}$. $K = \lambda fg\vec{h}\{e'\}^{IV}(f,g,\vec{h})$.

iv. Index for S^μ, where $\mu = (\sigma\to\tau\to\lambda)\to(\sigma\to\tau)\to\sigma\to\lambda$. $S(f,g,h) = f(h)(g(h))$ if $\lambda \neq 0$, $\simeq f(h)(g(h))$ if $\lambda = 0$. By lemma 55 there are indeces e_1, e_2 such that $\{e_1\}^{IV}(f,h,q,—) \simeq f(h,q,—)$, $\{e_2\}^{IV}(g,h,\vec{p}) \simeq g(h,\vec{p})$. We substitute $\lambda\vec{p}\{e_2\}^{IV}(g,h,\vec{p})$ for q in $\{e_1\}^{IV}(f,h,q,—)$. By lemma 54 and theorem 23 there is an index e such that $\{e\}^{IV}(f,g,h,—) \simeq \{e_1\}^{IV}(f,h,\lambda\vec{p}\{e_2\}^{IV}(g,h,\vec{p}),—)$. Then $\{e\}^{IV}(f,g,h,—) \simeq f(h,\lambda\vec{p}\{e_2\}^{IV}(g,h,\vec{p}),—) \simeq f(h,\lambda\vec{p}g(h,\vec{p}),—) \simeq f(h,g(h),—) \simeq f(h)(g(h))(—)$. Hence $S = \lambda fgh — \{e\}^{IV}(f,g,h,—)$.

v. Index for DC. $DC(x,y,f,g) = f$ if $x = y$, $= g$ if $x \neq y$. Using the scheme for primitive recursion (S 10) one can find an index e_1 such that $\{e_1\}^{IV}(0,f,g,z) \simeq f(z)$, $\{e_1\}^{IV}(n+1,f,g,z) \simeq g(z)$. The function t defined by $t(x,y) = 0$ if $x = y$, $= 1$ if $x \neq y$, is recursive in IV by S 3. Find e such that $\{e\}^{IV}(x,y,f,g,z) \simeq \{e_1\}^{IV}(t(x,y),f,g,z)$. Then $DC = \lambda xyfgz\{e\}^{IV}(x,y,f,g,z)$.

vi. Index for $FP^{(\tau\to\tau)\to\tau}$ where $\tau \neq 0$. By S 16, lemma 55 and S 12 there is an index e_1 such that $\{e_1\}^{IV}(x,—,g) \simeq g(\lambda\vec{p}\{x\}^{IV}(\vec{p},g),—)$ for all $x,—,g$, where $g \in HC(\tau\to\tau)$, and if $\{e_1\}^{IV}(x,—,g) \simeq t$ then there is an object $h \in HC(\tau)$ such that $h \leq \lambda\vec{p}\{x\}^{IV}(\vec{p},g)$, $g(h,—) \simeq t$, and $|\{e_1\}^{IV}(x,—,g)| > |\{x\}^{IV}(\vec{p},g)|$ when $h(\vec{p})\downarrow$. Choose e_2 such that $\{e_2\}^{IV}(y,—,g) \simeq \{e_1\}^{IV}(S^1(y,y),—,g)$, and if $\{e_1\}^{IV}(S^1(y,y),—,g)\downarrow$ then $|\{e_2\}^{IV}(y,—,g)| > |\{e_1\}^{IV}(S^1(y,y),—,g)|$. Let $x = S^1(e_2,e_2)$. Then $\{x\}^{IV}(—,g) \simeq \{S^1(e_2,e_2)\}^{IV}(—,g) \simeq \{e_2\}^{IV}(e_2,—,g) \simeq \{e_1\}^{IV}(S^1(e_2,e_2),—,g) \simeq \{e_1\}^{IV}(x,—,g) \simeq g(\lambda\vec{p}\{x\}^{IV}(\vec{p},g),—)$. Hence $\lambda\vec{p}\{x\}^{IV}(\vec{p},g)$ is a fixed point for g. Hence $FP(g) \leq \lambda\vec{p}\{x\}^{IV}(\vec{p},g)$ as $FP(g)$ is the least fixed point of g. To prove that $\lambda\vec{p}\{x\}^{IV}(\vec{p},g) \leq FP(g)$ we define $\{x\}_\mu$ as follows (μ is an ordinal):

$$\{x\}_\mu(\vec{p},g) \simeq \begin{cases} \{x\}^{IV}(\vec{p},g) & \text{if } |\{x\}^{IV}(\vec{p},g)| < \mu \\ \uparrow & \text{otherwise} \end{cases}$$

It is enough to prove that $\lambda\vec{p}\{x\}_\mu(\vec{p},g) \leq FP(g)$ for all ordinals μ. Obviously $\lambda\vec{p}\{x\}_o(\vec{p},g) \leq FP(g)$ as $\lambda\vec{p}\{x\}_o(\vec{p},g)$ is totally undefined. Suppose $\lambda\vec{p}\{x\}_\mu(\vec{p},g) \leq FP(g)$. In order to prove that $\lambda\vec{p}\{x\}_{\mu+1}(\vec{p},g) \leq FP(g)$ suppose $\{x\}_{\mu+1}(\text{—},g) \simeq t$. Then $\{x\}^{IV}(\text{—},g) \simeq t$, $|\{x\}^{IV}(\text{—},g)| \leq \mu$. Hence $\{S^1(e_2,e_2)\}^{IV}(\text{—},g) \simeq t$, $|\{S^1(e_2,e_2)\}^{IV}(\text{—},g)| \leq \mu$ as $x = S^1(e_2,e_2)$. By lemma 56 $\{e_2\}^{IV}(e_2,\text{—},g) \simeq t$, $|\{e_2\}^{IV}(\epsilon_2,\text{—},g)| < \mu$. By the construction of e_2 $\{e_1\}^{IV}(S^1(e_2,e_2),\text{—},g) \simeq t$ and $|\{e_1\}^{IV}(S^1(e_2,e_2),\text{—},g)| < \mu$. Hence $\{e_1\}^{IV}(x,\text{—},g) \simeq t$, $|\{e_1\}^{IV}(x,\text{—},g)| < \mu$. By the construction of e_1 $g(\lambda\vec{p}\{x\}^{IV}(\vec{p},g),\text{—}) \simeq t$, and there is an object $h \in HC(\tau)$ such that $h \leq \lambda\vec{p}\{x\}^{IV}(\vec{p},g)$, $g(h,\text{—}) \simeq t$, $|\{e_1\}^{IV}(x,\text{—},g)| > |\{x\}^{IV}(\vec{p},g)|$ for all $\vec{p}$ such that $h(\vec{p})\downarrow$. Hence $|\{x\}^{IV}(\vec{p},g)| < \mu$ when $h(\vec{p})\downarrow$. Hence $h \leq \lambda\vec{p}\{x\}_\mu(\vec{p},g)$. By the induction hypothesis $h \leq FP(g)$, hence $g(FP(g),\text{—}) \simeq t$ as $g(h,\text{—}) \simeq t$. Hence $FP(g)(\text{—}) \simeq t$, i.e. $\lambda\vec{p}\{x\}_{\mu+1}(\vec{p},\text{—}) \leq FP(g)$.

This proves that $\lambda\vec{p}\{x\}^{IV}(\vec{p},g) = FP(g)$ for all $g \in HC(\tau\to\tau)$, i.e. $FP = \lambda g\,\lambda\vec{p}\,\{x\}^{IV}(\vec{p},g)$.

vii. Suppose $f_1^{\sigma\to\tau} = \lambda g^\sigma \vec{h}\,\{e_1\}^{IV}(g^\sigma,\vec{h},h_1\ldots h_k)$, $f_2^\sigma = \lambda\vec{p}\{e_2\}^{IV}(\vec{p},h_1..h_k)$ ($f_2^\sigma \simeq \{e_2\}^{IV}(h_1\ldots h_k)$ if $\sigma = 0$). Substitute $\lambda\vec{p}\{e_2\}^{IV}(\vec{p},h_1\ldots h_k)$ for g^σ in $\{e_1\}^{IV}(g^\sigma,\vec{h},h_1\ldots h_k)$. By theorem 23 there is an index e such that $\{e\}^{IV}(\vec{h},h_1\ldots h_k) \simeq \{e_1\}^{IV}(\lambda\vec{p}\{e_2\}^{IV}(\vec{p},h_1\ldots h_k),\vec{h},h_1\ldots h_k)$. Hence $\{e\}^{IV}(\vec{h},h_1\ldots h_k) \simeq f_1^{\sigma\to\tau}(f_2^\sigma)(\vec{h})$, i.e. $f_1(f_2) = \lambda\vec{h}\{e\}^{IV}(\vec{h},h_1\ldots h_k)$.

□

To embed IV into $\mathcal{R}_\omega(\mathcal{B})$.

Lemma 57: Suppose $\mathcal{B}$ contain the hereditarily consistent objects M, K, L, O, 1. Then the functions u_j, d_j, u_j^{j+k}, d_j^{j+k}, M_k^s, $K_{k,j}^s$, Tr_μ^ν, $Intr_\mu^\nu$ are in $\mathcal{R}_o(\mathcal{B})$. If in addition $\mathcal{B}$ contains the characteristic function of N, +1 (= the successor function) and the function −1 (defined by: $0 \dot{-} 1 = 0$, $(k+1) \dot{-} 1 = k$, $x \dot{-} 1 = x$ if $x \notin N$), then each primitive recursive function (i.e. each function generated by S1 - S11, S15 - S16) is in $\mathcal{R}_1(\mathcal{B})$.

Proof: By lemma 49 each explicit definition is a combination, and hence is in $\mathcal{R}_o$, as each combination obviously is in $\mathcal{R}_o$. It is enough to prove that u_j, d_j, u_j^{j+k}, d_j^{j+k}, M_k^s, $K_{k,s}^s$, Tr_μ^ν, $Intr_\mu^\nu$ can be expressed as explicit definitions in which we allow occurrences of objects in $\mathcal{R}_o(\mathcal{B})$. If for instance T is an explicit definition in which $h_1,\dots,h_k$ occur, $h_1,\dots,h_k \in \mathcal{R}_o(\mathcal{B})$, let T' be the explicit definition $\lambda f_1,\dots,f_k\ T[\vec{h}/\vec{f}]$, where $T[\vec{h}/\vec{f}]$ is the expression we get by substituting $h_1,\dots,h_k$ by $f_1,\dots,f_k$ (assume that $f_1,\dots,f_k$ do not occur in T). Then $T' \in \mathcal{R}_o$. $T = T'h_1h_2 \dots h_k$, hence $T \in \mathcal{R}_o(\mathcal{B})$.

$u_o \in HC(0 \to 1)$ is defined by: $u_o(x)(y) = x$. Hence $u_o = K^{0\to 0\to 0}$. $d_o \in HC(1 \to 0)$ is defined by: $d_o(f) \simeq f(0)$. Hence $d_o = (\lambda y^0 \lambda f^1 . f(y))0$, and d_o is an explicit definition in which there is an occurrence of an object in $\mathcal{B}$, namely 0. Suppose $u_j, d_j \in \mathcal{R}_o(\mathcal{B})$. $u_{j+1} = \lambda f^{j+1} g^{j+1} . f(d_j g)$, $d_{j+1} = \lambda f^{j+2} g^j . f(u_j g^j)$. Hence u_{j+1}, d_{j+1} are explicit definitions in which there are occurrences of objects from $\mathcal{R}_o(\mathcal{B})$, namely u_j, d_j. $u_j^j = d_j^j = I^{j\to j}$, so $u_j^j, d_j^j \in \mathcal{R}_o$. Suppose u_j^{j+k}, $d_j^{j+k} \in \mathcal{R}_o(\mathcal{B})$. $u_j^{j+k+1} = \lambda f^j . u_{j+k}(u_j^{j+k} f^j)$, $d_j^{j+k+1} = \lambda f^{j+k+1} . d_j^{j+k}(d_{j+k} f)$, hence u_j^{j+k+1}, d_j^{j+k+1} are explicit definitions in which there are occurrences of objects from $\mathcal{R}_o(\mathcal{B})$.

$M_o^0 = 0 \in \mathcal{B}$. $M_1^0 = I^{0\to 0}$. $M_2^0 = M \in \mathcal{B}$. Suppose $M_{k+2}^0 \in \mathcal{R}_o(\mathcal{B})$. $M_{k+3}^0 = \lambda x_1,\dots,x_{k+3} . M(x_1, M_{k+2}^0(x_2 \dots x_{k+3}))$, hence M_{k+3}^0 is an explicit definition in which there are occurrences of objects from $\mathcal{R}_o(\mathcal{B})$. In the same way we see that $K_{k,j}^0 \in \mathcal{R}_o(\mathcal{B})$ when $1 \le j \le k$.

$M_o^{s+1} = \lambda g^s . 0$, $M_1^{s+1} = I^{s\to s}$, hence $M_o^{s+1}, M_1^{s+1} \in \mathcal{R}_o(\mathcal{B})$. Suppose $K_{2,1}^s, K_{2,2}^s \in \mathcal{R}_o(\mathcal{B})$.

$M_2^{s+1} = \lambda f^{s+1} g^{s+1} h^s [DC(d_o^s(K_{2,1}^s(h)), 0, u_o(f(K_{2,2}^s(h))), u_o(g(K_{2,2}^s(h))))0]$.

Hence M_2^{s+1} is an explicit definition in which there are occurrences of elements from $\mathcal{R}_o(\mathcal{B})$. One can easily prove that M_{k+3}^{s+1}, $K_{k,j}^{s+1}$,

$1 \leq j \leq k$, can be expressed as explicit definitions in which there are occurrences of elements from $\mathcal{R}_0(\mathcal{B})$.

The functions Tr_μ^ν, $Intr_\mu^\nu$ are defined inductively from similar functions of lower type, and by an easy induction one can prove that these functions are in $\mathcal{R}_0(\mathcal{B})$.

It remains to prove that each primitive recursive function is in $\mathcal{R}_1(\mathcal{B})$ when $\mathcal{B}$ in addition contains the characteristic function of N, the successor function and the function -1. This can be proved by induction on the schemes S1 - S11, S15 - S16. As examples we prove the result for S3 and S10.

S3: $f(x,y,—) = 0$ if $x = y$, $= 1$ if $x \neq y$. Hence $f(x,y,—) = DC(x,y,u_0(0),u_0(1))0$, i.e. $f = \lambda xy— .[DC(x,y,u_0(0),u_0(1))0]$, and $f \in \mathcal{R}_0(\mathcal{B})$.

S10: Suppose $\psi_1, \psi_2 \in \mathcal{R}_1(\mathcal{B})$, and let φ be defined by: $\varphi(0,—) \simeq \psi_1(—)$, $\varphi(n+1,—) \simeq \psi_2(\varphi(n,—),n,—)$ in $n \in N$, $\varphi(x,—) \simeq 0$ if $x \notin N$. Let $g \in HC$ be defined by: $g(f^1,0,—) \simeq \psi_1(—)$, $g(f^1,n+1,—) \simeq \psi_2(f(n),n,—)$, $g(f^1,x,—) \simeq 0$ if $x \notin N$. Let $p(—) = FP^{(1\to 1)\to 1}(\lambda fx.g(f,x,—))$. Then $\varphi(x,—) \simeq g(\lambda y\, p(—)(y),x,—)$. To prove that $\varphi \in \mathcal{R}_1(\mathcal{B})$ it suffices to prove that $g \in \mathcal{R}_1(\mathcal{B})$, because φ is an explicit definition built up from g and functions in $\mathcal{R}_1(\mathcal{B})$. The computation of $g(f,x,—)$ can be illustrated by the following diagram:

Input $f,x,—$.

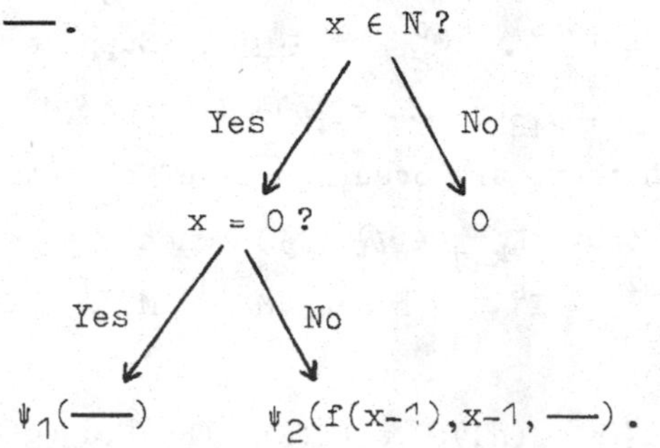

$g(f,x,—) \simeq DC(0,t(x),g_1,g_2)(0)$, where t is the characteristic function of N, $g_1 = DC(x,0,u_o(\psi_1(\text{---})),\ u_o(\psi_2(f(x-1),x-1,\text{---})))$, $g_2(x) = 0$ for all $x \in Ob$. We see that g can be expressed as an explicit definition in which the following objects from $\mathcal{R}_1(\mathcal{B})$ occur: $DC, 0, t, u_o, \psi_1, \psi_2, -1$. □

By iterated applications of DC one can prove

Lemma 58: Suppose $0 \in \mathcal{R}_1(\mathcal{B})$, $l \geq 0$. Suppose $\Psi_1,\ldots,\Psi_k \in \mathcal{R}_1(\mathcal{B})$. Suppose $R_1,\ldots,R_k$ are relations in $\mathcal{R}_1(\mathcal{B})$ (i.e. the characteristic functions of $R_1,\ldots,R_k$ are in $\mathcal{R}_1(\mathcal{B})$) such that for each list — exactly one $R_i(\text{---})$ is true. Let φ be the partial function defined by:

$$\begin{aligned}\varphi(\text{---}) &\simeq \psi_1(\text{---}) \quad \text{if } R_1(\text{---})\\ &\simeq \psi_2(\text{---}) \quad \text{if } R_2(\text{---})\\ &\ \vdots\\ &\simeq \psi_k(\text{---}) \quad \text{if } R_k(\text{---}).\end{aligned}$$

Then $\varphi \in \mathcal{R}_1(\mathcal{B})$.

Let $\vec{f} = f_1,\ldots,f_k$ be a sequence of objects in HC of types $\tau_1,\ldots,\tau_k$ such that $l(\tau_i) = l+1$ or $l(\tau_i) = l+2$ for $1 \leq i \leq k$. We will define an object $\Gamma \in HC(\tau_1 \to \tau_2 \to \ldots \to \tau_k \to l+1 \to l+1)$ such that $f^* = FP^{(l+1\to l+1)\to l+1}(\Gamma(\vec{f}))$ is the function which enumerates all computations $\{e\}^{IV}(\text{---})$ where each object in — is either of a type with level $\leq l$, or is in the list $\vec{f}$. f^* will satisfy the equation: $f^*(M^l(j,e,\vec{g})) \simeq \{e\}^{IV}(\text{---})$ where $\vec{g}$ is the list of objects in — of a type with level $\leq l$, the number j tells which objects from $\vec{f}$ occur in —.

Theorem 25: Suppose $\mathcal{B}$ contain the following objects: $M, K, L, 0, 1$, the characteristic function of N, the functions $+1$ and -1, $\vec{\varphi}, \mathcal{F}$, where $\vec{\varphi}, \mathcal{F}$ are the hereditarily consistent objects in the schemes

S 13 and S 14. Then there is an object $\Gamma \in \mathcal{R}_1(\mathcal{B})$ with the properties mentioned above.

Corollary: Suppose $\{e\}^{IV}$ is of type τ with $l(\tau) \leq l+3$. Then $\{e\}^{IV} \in \mathcal{R}_{l+1}(\mathcal{B})$.

Proof of the corollary: Let $\tau_1,\ldots,\tau_k$ be the types of level $l+1$ and $l+2$ in the argument list of $\{e\}^{IV}$. Let —— be a sequence with the same signature as the argument list of $\{e\}^{IV}$. Let $\vec{g}$ be the list of those objects in —— which is of a type with level $\leq l$, let $\vec{f}$ be the list of those objects in —— which are of types with level $l+1$ or $l+2$. Then $\{e\}^{IV}(\text{——}) \simeq FP^{(l+1\to l+1)\to l+1}(\Gamma(\vec{f}))M^l(j,e,\vec{g})$, where j is the number which specifies $\vec{f}$. Now $\Gamma \in \mathcal{R}_1(\mathcal{B})$ by the theorem. $\lambda je\vec{g}M^l(j,e,\vec{g}) \in \mathcal{R}_o(\mathcal{B})$ when we fix the signature of $\vec{g}$. Hence $\{e\}^{IV} \in \mathcal{R}_{l+1}(\mathcal{B})$. $\square$

Proof of theorem 25: We will define Γ by cases. Let $\vec{f} = f_1,\ldots,f_k$, $f_i \in HC(\tau_i)$, $l(\tau_i) = l+1$ or $l+2$. We will define $\Gamma(\vec{f},f^{l+1},p^l)$. p^l will be regarded as a sequence $M^l(j,e,g_1 \ldots g_r)$, where j, e are natural numbers, $g_1,\ldots,g_r$ are hereditarily consistent objects of types with level $\leq l$. Suppose $\{e\}^{IV}(\text{——})\downarrow$ and $\{e'\}^{IV}(\text{——}')$ is a subcomputation of $\{e\}^{IV}(\text{——})$, and the level of the type of each element in —— is $\leq l+2$. If there is an element of a type of level $>l$ in ——' then that element is in ——. This can easily be proved by inspection of the schemes. The number j in $M^l(j,e,g_1 \ldots g_r)$ will tell us which objects from $\vec{f}$ will occur in the argument list of $\{e\}^{IV}$, and how they are distributed in that list. If p^l codes a convergent computation relative to some objects from $\vec{f}$ then each subcomputation will be coded by another p^l.

To define $\Gamma(\vec{f},f^{l+1},p^l)$ we first find j and l ($= K_1^l(p)$, $K_2^l(p)$ respectively). From e we can find the scheme for the computation $\{e\}^{IV}(\text{——})$, where —— is the list made up from $g_1,\ldots,g_r$ ($= K_3^l(p)$,..

$\ldots, K^1_{3+r}(p)$ respectively) and objects from $\vec{f}$ as indicated by j and e. There is one case for each scheme, and one otherwise case (which applies when we can immediately see that e is not an index for a convergent computation, or if the signature of the list —— is not the same as the signature of the argument list of $\{e\}^{IV}$. There are primitive recursive relations $R_o, \ldots, R_{16}$ such that for each p^l exactly one $R_i(p^l)$ is true. R_o is for the otherwise case, R_1 for $S1, \ldots, R_{16}$ for S16. We let

$$\begin{aligned} \Gamma(\vec{f}, f^{l+1}, p^l) &\simeq \psi_o(\vec{f}, f, p) \quad \text{if} \quad R_o(p) \\ &\simeq \psi_1(\vec{f}, f, p) \quad \text{if} \quad R_1(p) \\ &\vdots \\ &\simeq \psi_{16}(\vec{f}, f, p) \quad \text{if} \quad R_{16}(p). \end{aligned}$$

We define ψ_o, ψ_3, ψ_9, ψ_{12}, ψ_{14} and ψ_{16} as examples.

$\psi_o(\vec{f}, f, p) \simeq f(p) + 1$ (in order to make $FP(\Gamma(\vec{f}))(p)$ divergent. For $FP(\Gamma(\vec{f}))(p) \simeq \Gamma(\vec{f}, FP(\Gamma(f)), p) \simeq FP(\Gamma(\vec{f}))(p) + 1$, hence $FP(\Gamma(\vec{f}))(p)\uparrow$.)

ψ_3: Now $p = M^l(j, \langle 3, n\rangle, x, y, \vec{g})$. Let $\psi_3(\vec{f}, f, p) \simeq t(x, y)$, where t is the initial function of S3.

ψ_9: Now $p = M^l(j, \langle 9, n, e_1, e_2\rangle, \vec{g})$. Let $\psi_9(\vec{f}, f, p) \simeq f(M^l(j', f(M^l(j, e_2, \vec{g})), \vec{g}))$, where j' specifies the distribution of objects from $\vec{f}$ in the argument list of $\{e_1\}^{IV}$.

ψ_{12}: Now $p = M^l(j, \langle 12, n\rangle, x, \vec{g})$. Let $\psi_{12}(\vec{f}, f, p) \simeq f(M^l(j', x, \vec{g}))$, where j' specifies the distribution of objects from $\vec{f}$ in the argument list of $\{x\}^{IV}$.

ψ_{14}: Now $p = M^l(j, \langle 14, n, i, \langle e_1 \ldots e_t\rangle\rangle, \vec{g})$. Let $\psi_{14}(\vec{f}, f, p) \simeq \mathcal{F}_i(\lambda\vec{x}\, f(M^l(j_1, e_1, \vec{x}, \vec{g})), \ldots, \lambda\vec{x}\, f(M^l(j_t, e_t, \vec{x}, \vec{g})))$, where j_s specifies the distribution of objects from $\vec{f}$ in the argument list of $\{e_s\}^{IV}$.

ψ_{16}: Now $p = M^l(j, \langle 16, n, \langle e_1 \ldots e_t\rangle\rangle, \vec{g})$. Let $\psi_{16}(\vec{f}, f, p) \simeq p^{\tau}(\lambda\vec{q}\, f(M^l(j_1, e_1, \vec{q}, \vec{g})), \ldots, \lambda\vec{q}\, f(M^l(j_t, e_t, \vec{q}, \vec{g})))$, where j_s specifies the distribution of objects from $\vec{f}$ in the argument list of $\{e_s\}^{IV}$. Note that the level of the type of an element in a list $\vec{q}$ is at most l. p^τ is either in the list $\vec{g}$ or in the list $\vec{f}$.

We see that each ψ_i can be expressed as explicit definitions in which there are occurrances of objects from $\mathcal{R}_1(\mathcal{B})$. Hence $\psi_i \in \mathcal{R}_1(\mathcal{B})$. The characteristic functions of $R_0,\dots,R_{16}$ are primitive recursive, and hence in $\mathcal{R}_1(\mathcal{B})$ by lemma 57. By lemma 58 $\Gamma \in \mathcal{R}_1(\mathcal{B})$.

Let $f^* = FP^{(l+1\to l+1)\to l+1}(\Gamma(\vec{f}))$. $f^* \in \mathcal{R}_{l+1}(\mathcal{B})$. Let $f_0 = \uparrow^{l+1}$ (= the totally undefined object of type $l+1$), $f_{\beta+1} = \Gamma(\vec{f})(f_\beta)$, $f_\lambda = \bigcup_{\beta<\lambda} f_\beta$ if λ is a limit ordinal. Then $f^* = \bigcup\{f_\beta : \beta$ is an ordinal$\}$. By induction on λ one can prove that if $f_\lambda(p) \simeq t$ then $p = M^l(j,e,\vec{g})$ and $\{e\}^{IV}(\text{——}) \simeq t$, where —— is the sequence of objects from $\vec{g},\vec{f}$ indicated by j and e. By induction on $|\{e\}^{IV}(\text{——})|$ one can prove that if $\{e\}^{IV}(\text{——}) \simeq t$, and —— consists of objects of types with level $\leq l$ ($=\vec{g}$), and objects from $\vec{f}$, then $f^*(M^l(j,e,\vec{g})) \simeq t$, where j is the number which specifies the objects from $\vec{f}$ in ——. Both inductions are straightforward. □

The main results.

If $X \subseteq HC$ let $X^l = \{f^\tau : f \in X, l(\tau) \leq l\}$.

Theorem 26 (the reduction theorem): Suppose $\mathcal{B}$ contain the hereditarily consistent objects $M, K, L, 0, 1$, the characteristic function of N, the functions $+1$, -1, and $\mathcal{B} \subseteq HC^{l+2}$. Then $\mathcal{R}_\omega(\mathcal{B})^{l+3} = \mathcal{R}_{l+1}(\mathcal{B})^{l+3}$.

Proof: Obviously $\mathcal{R}_{l+1}(\mathcal{B})^{l+3} \subseteq \mathcal{R}_\omega(\mathcal{B})^{l+3}$. To prove the opposite inclusion suppose $f \in \mathcal{R}_\omega(\mathcal{B})^{l+3}$. Then there are $h_1,\dots,h_k \in \mathcal{B}$ such that $f \in \mathcal{R}_\omega(\{h_1 \dots h_k\})^{l+3}$. By theorem 24 there is an index e such that $f = \lambda\vec{g}\{e\}^{IV}(\vec{g},h_1 \dots h_k)$. The level of the type of an element in $\vec{g},h_1 \dots h_k$ is $\leq l+2$ because $\mathcal{B} \subseteq HC^{l+2}$ and the level of the type of f is $\leq l+3$. By the corollary of theorem 25 $\{e\}^{IV} \in \mathcal{R}_{l+1}(\mathcal{B})$. Hence $f \in \mathcal{R}_{l+1}(\mathcal{B})$, as $f = \lambda\vec{g}.(\{e\}^{IV})\vec{g}\, h_1 \dots h_k$. □

Remark: Platek has proved the reduction theorem in [26] (theorem 5.3.2a page 165). His proof involves λ-calculus. In the proof of theorem 26 the main idea is to introduce the indeces and schemes of IV, and then go back and forth between IV and $\mathcal{R}_\omega$ by theorems 24 and 25.

Theorem 27: Let $\mathcal{B}$ be the following list of hereditarily consistent objects: M, K, L, 0, 1, the characteristic function of N, the functions +1, -1, $\vec{\varphi}$, $\mathcal{F}$.

i) Suppose $\vec{x}$ is a list of variables of type 0. Let $\vec{\gamma}$ be a list of objects such that the level of the type of each object is $\leq l+2$. Then for all e: $\Psi(\lambda\vec{x}\{e\}^{III}(\vec{x},\vec{\gamma})) \in \mathcal{R}_{l+1}(\mathcal{B} \cup \{\Psi\vec{\gamma}\})$.

ii) Suppose $\vec{\beta}$ is a list of variables such that the level of the type of each variable $\leq l+1$. Suppose $\vec{\gamma}$ is as in i), and that there is an object in $\vec{\gamma}$ of a type $l+2$. Then for all e:

$$\Psi(\lambda\vec{\beta}\{e\}^{III}(\vec{\beta},\vec{\gamma})) \in \mathcal{R}_{l+1}(\mathcal{B} \cup \{\Psi\vec{\gamma}\}).$$

iii) Let $\vec{\gamma}$ be any sequence of total objects. If $f \in \mathcal{R}_\omega(\mathcal{B} \cup \{\Psi\vec{\gamma}\})$ then Φf is partial recursive in $\vec{\gamma}$ in theory III.

Corollary: Let φ be a partial function of a type with level 1. Then φ is partial recursive in I iff $\varphi \in \mathcal{R}_1(\mathcal{B})$.

Proof of corollary: => follows from i) in the theorem (with the empty sequence $\vec{\gamma}$, $l=0$) and the lemmas 50, 51. <= follows by iii) and the lemmas 50, 51. □.

Proof of theorem 27: i) and ii). Suppose $\vec{\gamma}$ is as in i), and $\vec{\beta}$ is a list of variables of types with level $\leq l+1$. By lemma 53 $\{e\}^{III}(\vec{\beta},\vec{\gamma}) \simeq \{e\}^{IV}(\Psi\vec{\beta},\Psi\vec{\gamma})$ for all $\vec{\beta},\vec{\gamma}$. $\Psi(\lambda\vec{\beta}\{e\}^{III}(\vec{\beta},\vec{\gamma}))(\vec{g}) \simeq \{e\}^{III}(\Phi\vec{g},\vec{\gamma})$

$$\simeq \begin{cases} \uparrow & \text{if } \Phi g \text{ is not total for some } g \text{ in } \vec{g} \\ \{e\}^{III}(\vec{\beta},\vec{\gamma}) & \text{if } \Phi\vec{g} = \vec{\beta}. \end{cases}$$

$\{e\}^{IV}(\vec{g},\Psi\vec{\gamma})$ may be defined even if Φg is not total for some g in $\vec{g}$.

Hence $\Psi(\lambda\vec{\beta}\{e\}^{III}(\vec{\beta},\vec{\gamma})) \leq \lambda\vec{g}\{e\}^{IV}(g,\Psi\vec{\gamma})$. In i) $\Psi(\lambda\vec{x}\{e\}^{III}(\vec{x},\vec{\gamma})) = \lambda\vec{x}\{e\}^{IV}(\vec{x},\Psi\vec{\gamma})$, as the problem of totality is not present at level 0. Hence $\Psi(\lambda\vec{x}\{e\}^{III}(\vec{x},\vec{\gamma})) \in \mathcal{R}_{l+1}(\mathcal{B} \cup \{\Psi\vec{\gamma}\})$ because $\lambda\vec{x}\{e\}^{IV}(\vec{x},\Psi\vec{\gamma}) \in \mathcal{R}_{l+1}(\mathcal{B} \cup \{\Psi\vec{\gamma}\})$ by the corollary of theorem 25.

Suppose $\vec{\gamma}$ satisfies the conditions of ii). To prove that $\Psi(\lambda\vec{\beta}\{e\}^{III}(\vec{\beta},\vec{\gamma})) \in \mathcal{R}_{l+1}(\mathcal{B} \cup \{\Psi\vec{\gamma}\})$ it suffices to find an index e' such that $\Psi(\lambda\vec{\beta}\{e\}^{III}(\vec{\beta},\vec{\gamma})) = \lambda\vec{g}\{e'\}^{IV}(\vec{g},\Psi\vec{\gamma})$. Then ii) follows by the corollary of theorem 25. e' will be an index such that

$$\{e'\}^{IV}(\vec{g},\Psi\vec{\gamma}) \simeq \begin{cases} \uparrow & \text{if } \Phi g \text{ is not total for some } g \text{ in } \vec{g} \\ \{e\}^{IV}(\vec{g},\Psi\vec{\gamma}) & \text{if each } \Phi g \text{ is total.} \end{cases}$$

It follows from this that $\Psi(\lambda\vec{\beta}\{e\}^{III}(\vec{\beta},\vec{\gamma})) = \lambda\vec{g}\{e'\}^{IV}(\vec{g},\Psi\vec{\gamma})$.

Suppose g is in the list $\vec{g}$. We check whether or not g is total in the following way: Let τ be the type of g, $m = l(\tau)$. Then $m \leq l+1$. Choose a γ from the list $\vec{\gamma}$ such that $l(\sigma) = l+2$, where σ is the type of γ. By looking at the definitions of $\dot{u}_m^{l+1}$ and Tr_τ^m one can prove that Φg is total iff $\Phi\, u_m^{l+1}(Tr_\tau^m(g))$ is total. Now $\Phi\, u_m^{l+1}(Tr_\tau^m(g))$ is total iff $(\Psi\gamma)(u_m^{l+1}(Tr_\tau^m(g)))\downarrow$, hence Φg is total iff $(\Psi\gamma)(u_m^{l+1}(Tr_\tau^m(g)))\downarrow$. By lemma 57 u_m^{l+1} and Tr_τ^m are in $\mathcal{R}_0(\mathcal{B})$. By theorem 24 they are recursive in IV. We can let $\{e'\}^{IV}(\vec{g},\Psi\vec{\gamma})$ be the computation which first computes $(\Psi\gamma)(u_m^{l+1}(Tr_\tau^m(g)))$ for all g in $\vec{g}$. If these computations converges the output is $\{e\}^{IV}(\vec{g},\Psi\vec{\gamma})$.

iii) Suppose $f \in \mathcal{R}_\omega(\mathcal{B} \cup \{\Psi\vec{\gamma}\})$. Then there are objects $h_1,\ldots,h_k \in \mathcal{B}$ such that $f \in \mathcal{R}_\omega(\{h_1 \ldots h_k,\Psi\vec{\gamma}\})$. By theorem 24 there is an index e such that $f = \lambda\vec{g}\{e\}^{IV}(\vec{g},h_1 \ldots h_k,\Psi\vec{\gamma})$. Now each object in $\mathcal{B}$ is partial recursive in IV. By theorem 23 we can substitute for $h_1,\ldots,h_k$ and find an index e' such that $f = \lambda\vec{g}\{e'\}^{IV}(\vec{g},\Psi\vec{\gamma})$. $\Phi(f)(\vec{\beta}) \simeq f(\Psi\vec{\beta}) \simeq \{e'\}^{IV}(\Psi\vec{\beta},\Psi\vec{\gamma}) \simeq \{e'\}^{III}(\vec{\beta},\vec{\gamma})$ (the last equality by lemma 53). Hence $\Phi(f) = \lambda\vec{\beta}\{e'\}^{III}(\vec{\beta},\vec{\gamma})$, i.e. Φf is partial recursive in $\vec{\gamma}$ in III. $\square$

An indexfree representation of a recursion theory on two types.

Let $\mathfrak{A}$ be the computation domain $(I, S, \mathcal{S})$ (cf. § 1). Let $\mathcal{L}$ be the list $R_1,\dots,R_k, \varphi_1,\dots,\varphi_l, F_1,\dots,F_m$ of predicates, partial functions and functionals. We can adapt this recursion theory to the setting of this chapter in the following way.

Let $Ob = S$. Now $I = S \cup S_\omega$. S_ω is a subset of $T(1)$ (= the total objects of type 1 over S). If φ is a partial function $I \to S$ then $\varphi \notin HC$ because $I \neq HC(\tau)$. But there are two partial objects which are naturally associated to φ: $\varphi_1 \in PT(1)$, $\varphi_2 \in PT(2)$, where $\varphi_1(s) \simeq \varphi(s)$ if $s \in S$, $\varphi_2(\alpha^1) \simeq \varphi(\alpha)$ if $\alpha \in {}^S\omega$, $\simeq \uparrow$ if $\alpha^1 \notin {}^S\omega$. If φ is a partial function $I^j \to S$, where $j \geq 1$, then there are 2^j partial objects associated to φ. If for instance $j = 2$ we get $\varphi_1 \in PT(0\to0\to0)$, $\varphi_2 \in PT(1\to0\to0)$, $\varphi_3 \in PT(0\to1\to0)$, $\varphi_4 \in PT(1\to1\to0)$, where $\varphi_1(x,y) \simeq \varphi(x,y)$ if $x,y \in S$, $\varphi_2(\alpha^1,x) \simeq \varphi(\alpha,x)$ if $\alpha \in {}^S\omega$, $x \in S$, $\simeq \uparrow$ if $\alpha \notin S_\omega$.
$\varphi_3(x,\alpha^1) \simeq \varphi(x,\alpha^1)$ if $x \in S$, $\alpha^1 \in {}^S\omega$, $\simeq \uparrow$ if $\alpha \notin {}^S\omega$.
$\varphi_4(\alpha^1,\beta^1) \simeq \varphi(\alpha^1,\beta^1)$ if $\alpha^1,\beta^1 \in {}^S\omega$, $\simeq \uparrow$ otherwise.

Suppose F is a functional as defined in § 2, i.e. F is a total function ${}^IS \to \omega$. We can associate an $F' \in T(1\to2\to0)$ as follows: $F'(\alpha^1,\alpha^2) = F(\beta)$, where $\beta \in {}^IS$ is defined by: $\beta(s) = \alpha^1(s)$ if $s \in S$, $\beta(x) = \alpha^2(x)$ if $x \in {}^S\omega$.

Let $\mathcal{L}'$ be the list of partial and total objects associated to the objects in $\mathcal{L}$. The partial objects stem from the φ's, and have types with level 1 or 2. We let $\mathbb{III}$ be the recursion theory built up from these partial objects (i.e. these are the objects of S 13 and S 14). The following result can be proved by induction on the lengths of the computations:

Lemma 59: i) There is a primitive recursive function δ_3 such that for all e,i $\lambda\vec{\alpha}\{\delta_3(e,i)\}^{\mathbb{III}}(\vec{\alpha}, —)$ is the i-th partial function associated to $\{e\}^{\mathcal{L}}$ as mentioned above. The list — consists of objects associated to the functionals.

ii) Suppose that the equality relation on S is recursive in $\mathcal{L}$. Suppose φ is a partial function of a type with level ≤ 2, each type in the argument list of φ is pure, and φ is partial recursive in —— in III, where —— is as in i). Then the restriction of φ to ${}^{S}\omega$ is partial recursive in $\mathcal{L}$ (where the restriction of φ is defined by: at each argument place of φ of type 1 we allow objects from ${}^{S}\omega$).

Combining this with theorem 27 we get

<u>Theorem 28</u>: Let $\mathcal{B}$ be the following list of hereditarily consistent objects: M, K, L, O, 1, the characteristic function of N, the functions +1, -1, $\Psi\mathcal{L}'$.

i) If φ is partial recursive in $\mathcal{L}$, and there is at least one total object in $\mathcal{L}$ then $\Psi\varphi_i \in \mathcal{R}_2(\mathcal{B})$, where φ_i is the i-th partial function associated to φ.

ii) If there is no total object in $\mathcal{L}$, and φ, φ_i are as in i), then $\Psi(\varphi_i) \in \mathcal{R}_1(\mathcal{B} \cup \Psi(\alpha^2))$, where $\alpha^2(\beta^1) = 0$ if $\beta \in {}^{S}\omega$, $= 1$ if not.

iii) If $f \in \mathcal{R}_\omega(\mathcal{B})$ and the type of each argument place in f is pure (i.e. the type of f is special), and the level of the type of f is ≤ 2 then the restriction of Φf to ${}^{S}\omega$ is partial recursive in $\mathcal{L}$.

Remark: In i) we need the total object in $\mathcal{L}$ to decide if objects in HC(1) are total and in ${}^{S}\omega$. The associates to the functionals in $\mathcal{L}$ are of types of level 3. So $\Psi\varphi_i \in \mathcal{R}_k(\mathcal{B})$ where $k = 3-1 = 2$. In ii) α^2 plays the role of the total object in i). There are no functionals as the list $\mathcal{L}$ does not contain any total object, hence no object of a type of level 3. Hence the maximum level of an associate is 2, and $\Psi(\varphi_i) \in \mathcal{R}_k(\mathcal{B} \cup \Psi(\alpha^2))$, where $k = 2-1 = 1$.

<u>A codingfree proof of the reduction theorem.</u>

We will end this chapter by a direct proof of the reduction theorem.

The proof depends neither on λ-calculus as theorem 5.3.2 in [26] nor on indexing as theorem 26 in this chapter. It turns out that we can weaken the assumptions about $\mathcal{B}$. It is not necessary to assume that $\mathcal{B}$ contains the objects M, K, L, O, 1 etc. as we do in theorem 26.

Theorem 29: Suppose $\mathcal{B} \subseteq HC^{l+2}$. If $l > 0$ then $\mathcal{R}_\omega(\mathcal{B})^{l+3} = \mathcal{R}_{l+1}(\mathcal{B})^{l+3}$. If $\uparrow \in \mathcal{R}_1(\mathcal{B})$ then $\mathcal{R}_\omega(\mathcal{B})^3 = \mathcal{R}_1(\mathcal{B})^3$. ($\uparrow$ is "being undefined". By conventions made earlier in this chapter we can deal with $\uparrow$ as an object of type 0. $\uparrow \leq x$ for all $x \in Ob$. If $f \in HC(0 \to \tau)$ then $f(\uparrow) = \uparrow^\tau$, where $\uparrow^\tau$ is the hereditarily consistent object of type τ which is totally undefined if $\tau \neq 0$, $\uparrow^0 = \uparrow$.)

Proof: We prove the theorem by 4 propositions.

Proposition 1: Suppose that $\mathcal{R}_{m+2}(\mathcal{C})^{m+3} = \mathcal{R}_{m+1}(\mathcal{C})^{m+3}$ whenever $\mathcal{C} \subseteq HC^{m+2}$, $m > 0$, and $\mathcal{R}_2(\mathcal{C})^3 = \mathcal{R}_1(\mathcal{C})^3$ whenever $\mathcal{C} \subseteq HC^2$, $\uparrow \in \mathcal{R}_1(\mathcal{C})$. Then $\mathcal{R}_\omega(\mathcal{B})^{l+3} = \mathcal{R}_{l+1}(\mathcal{B})^{l+3}$ when $\mathcal{B}$ is as in theorem 29.

Proof: To prove that $\mathcal{R}_\omega(\mathcal{B})^{l+3} \subseteq \mathcal{R}_{l+1}(\mathcal{B})^{l+3}$ suppose $f \in \mathcal{R}_\omega(\mathcal{B})^{l+3}$. Choose $m \geq l$ such that $f \in \mathcal{R}_{m+2}(\mathcal{B})^{l+3}$. Let $\mathcal{C} = \mathcal{B}$. Then $\mathcal{C} \subseteq HC^{m+2}$. By the assumption $f \in \mathcal{R}_{m+1}(\mathcal{B})^{l+3}$. Repeating this with m-1, m-2, ..., l for m we finally get $f \in \mathcal{R}_{l+1}(\mathcal{B})^{l+3}$. □ (prop. 1)

So it suffices to prove that $\mathcal{R}_{l+2}(\mathcal{B})^{l+3} = \mathcal{R}_{l+1}(\mathcal{B})^{l+3}$ under the assumptions of theorem 29.

Proposition 2: $\uparrow^\tau \in \mathcal{R}_{l+1}(\mathcal{B})$ for all type symbols τ.

Proof: It suffices to prove that $\uparrow \in \mathcal{R}_{l+1}(\mathcal{B})$. For if $\tau \neq 0$ then $\uparrow^\tau = (\lambda x^0 g_1 \ldots g_k . x^0)\uparrow$, where g_i is a variable for $HC(\tau_i)$, $\tau_1 \ldots \tau_k$ are the factors of τ. $\lambda x^0 g_1 \ldots g_k . x^0$ is an explicit definition, hence a combination by lemma 49. Each combination is in $\mathcal{R}_0$. So if $\uparrow \in \mathcal{R}_{l+1}(\mathcal{B})$ it follows that $\uparrow^\tau \in \mathcal{R}_{l+1}(\mathcal{B})$.

If $l = 0$ then $\uparrow \in \mathcal{R}_{l+1}(\mathcal{B})$ by assumption. If $l > 0$ then

$FP^{(2\to 2)\to 2}I^{2\to 2}$ and $FP^{(1\to 1)\to 1}I^{1\to 1}$ ($= \uparrow^2, \uparrow^1$ respectively) are in $\mathcal{R}_2(\mathcal{B})$, hence in $\mathcal{R}_{1+1}(\mathcal{B})$. Now $\uparrow = \uparrow^2\uparrow^1$, hence $\uparrow \in \mathcal{R}_{1+1}(\mathcal{B})$.

□ (prop. 2)

<u>Observation</u>: $\uparrow^{(o\to o)\to o}$ can be regarded as $FP^{(o\to o)\to o}$. For $\uparrow^{(o\to o)\to o}f^1 = \uparrow$, and $f^1\uparrow = \uparrow$, i.e. $\uparrow$ is a fixed point for f^1. Obviously $\uparrow$ is the least fixed point for f^1. Let $p^o = \uparrow$, $p^{\alpha+1} = f^1p^\alpha$, $p^\lambda = \bigcup_{\beta<\lambda} p^\beta$ if λ is a limit ordinal, $p = \bigcup\{p^\alpha : \alpha \in On\}$. Then we easily see that $p^\alpha = \uparrow$ for all ordinals α, and $p = \uparrow$. Hence p is the least fixed point for f^1.

Remark: The observation above is based on the following convention: $f\uparrow = \uparrow^\tau$ when $f \in HC(0\to\tau)$. Suppose we do not make this convention, i.e. $f(\uparrow)$ can be any object in $HC(\tau)$. Nevertheless there is a least fixed point for each f in $HC(1)$. This least fixed point is $f\uparrow$. For if $f\uparrow = \uparrow$ then $f(f\uparrow) = f\uparrow$. If $f\uparrow = x$ where $x \in Ob$ then $\uparrow \leq x$. Hence $f\uparrow \leq fx$ as $f \in HC$. Hence $x \leq fx$. This is possible only when $fx = x$, as $x \not\leq y$ when $x \neq y$, $x,y \in Ob$, and $x \not\leq \uparrow$. Hence $f(f\uparrow) = f\uparrow$. Let $p^o = \uparrow$, $p^{\alpha+1} = fp^\alpha$, $p^\lambda = \bigcup_{\beta<\lambda} p^\beta$ if λ is a limit ordinal, $p = \bigcup\{p^\alpha : \alpha \in On\}$. Then $p = \uparrow$ if $f\uparrow = \uparrow$, $p = x$ if $f\uparrow = x$. Hence p is the least fixed point of f. We can let $FP^{(o\to o)\to o} = \lambda f^1 . f\uparrow$.

By proposition 2 $FP^{(o\to o)\to o} \in \mathcal{R}_{1+1}(\mathcal{B})$ regardless of the convention we make about $f\uparrow$.

If $f \in \mathcal{R}_\omega(\mathcal{B})$ then $f = f_1f_2 \dots f_n$, where f_1 is an initial function, and $f_2,\dots,f_n$ may be composite. An initial function p <u>occurs</u> in f if $p = f_1$ or if p occurs in an f_i, $2 \leq i \leq n$.

<u>Proposition 3</u>: Each f in $\mathcal{R}_{1+2}(\mathcal{B})^{1+3}$ can be expressed as an explicit definition in which there is no occurrence of a combinator (i.e. the objects I, K, S). The other initial objects may occur in the explicit

definition. If ν is the type of a variable in the explicit definition then $l(\nu) \leq l+2$.

Proof: Suppose $f \in \mathcal{R}_{l+2}(\mathcal{B})^{l+3}$. Then $f = f_1 f_2 \ldots f_n$ where f_1 is initial. We will eliminate the combinators from the left until the leftmost object is an initial function which is not a combinator. If $f_1 = I$ and $n \geq 2$ then $f = If_2 \ldots f_n = f_2 \ldots f_n$. If $n = 1$ then $f = I = \lambda v . v$. If $f_1 = K$ and $n \geq 3$ then $f = Kf_2 f_3 \ldots f_n = f_2 f_4 \ldots$ $\ldots f_n$. If $n = 2$ then $f = Kf_2 = \lambda v . Kf_2 v = \lambda v . f_2$. If $n = 1$ then $f = K = \lambda v_1 v_2 . v_1$. If $f_1 = S$ and $n \geq 4$ then $f = Sf_2 f_3 f_4 \ldots f_n = f_2 f_4 (f_3 f_4) \ldots f_n$. If $n = 3$ then $f = Sf_2 f_3 = \lambda v . Sf_2 f_3 v = \lambda v . f_2 v (f_3 v)$. If $n = 2$ then $f = Sf_2 = \lambda v_1 v_2 . Sf_2 v_1 v_2 = \lambda v_1 v_2 . f_2 v_2 (v_1 v_2)$. If $n = 1$ then $f = S = \lambda v_1 v_2 v_3 . v_1 v_3 (v_2 v_3)$. After a finite number of eliminations we have the following expression for f: $f = \lambda \vec{v} . q q_1 \ldots q_m$ where q is initial and not a combinator, and $q_1 \ldots q_m$ may be composite. Then we perform the same eliminations for $q_1, q_2 \ldots q_m$, and so on until there is no combinator left in the expression for f.

Note that the level of the type of each object in $\vec{v}$ is $\leq l+2$ because these types are factors in μ (= the type of f) and $l(\mu) \leq l+3$. q is an element in the following set of objects:
$\mathcal{B} \cup \{\vec{v}\} \cup \{DC\} \cup \{FP^{(\tau\to\tau)\to\tau} : l(\tau) \leq l+2\}$.
In any case the level of the type of q is $\leq l+4$. Hence the level of the type of each q_i is $\leq l+3$, as for f. Hence the new variables introduced in $q_1, \ldots, q_m$ will also be of types with level $\leq l+2$, etc.

□ (prop. 3)

Theorem 29 follows from propositions 1 and 4.

<u>Proposition 4</u>: $\mathcal{R}_{l+2}(\mathcal{B})^{l+3} \subseteq \mathcal{R}_{l+1}(\mathcal{B})^{l+3}$.

Proof: Suppose $f \in \mathcal{R}_{l+2}(\mathcal{B})^{l+3}$. By proposition 3 f can be expressed as an explicit definition in which the combinators do not occur. In which way can an initial object $FP^{(\tau\to\tau)\to\tau}$ with $l(\tau) = l+2$ occur in the explicit definition? It is impossible that it can occur as $\lambda\vec{u}\,.\,p_1 p_2 \ldots p_k FP\, p_{k+2} \ldots p_n$ with $k \geq 1$ because then the type of p_1 must be of a level above the level of FP, hence above $1+4$. p_1 is also initial and there are no initial objects of a type with level above $1+4$ (the combinators are eliminated). Hence FP can occur only as $\lambda\vec{u}\,.\,FPp_2p_3 \ldots p_n$. It is impossible that $n = 1$. Then this expression would be $\lambda\vec{u}\,.\,FP$. The level of this is $1+4$. Either $\lambda\vec{u}.FP$ is an argument of an initial function, or it is f, and both cases are impossible by considering the level of the initial function or f. Hence FP can occur only as $\lambda\vec{u}\,.\,FPp_2p_3 \ldots p_n$ with $n \geq 2$.

Now we will eliminate each $FP^{(\tau\to\tau)\to\tau}$ with $l(\tau) = 1+2$ from the explicit definition. We begin to eliminate at the innermost places of the expression, i.e. first we eliminate FP in $\lambda\vec{u}\,.\,FPp_2 \ldots p_n$ where there is no occurrence of an $FP^{(\lambda\to\lambda)\to\lambda}$ with $l(\lambda) = 1+2$ in $p_2 \ldots p_n$. Then we eliminate the next FP in the same way, and so on.

Suppose we have an occurrence of an $FP^{(\tau\to\tau)\to\tau}$ with $l(\tau) = 1+2$ as just described. The type of p_2 is $\tau \to \tau$. p_2 can be expressed as $\lambda p\, \lambda\vec{v}\,.\,q\,q_1 \ldots q_r$, where the type of p is τ, the types of the variables in $\vec{v}$ are the factors of τ, q is an element in the following set of objects and variables: $\mathcal{B} \cup \{DC\} \cup \{FP^{(\lambda\to\lambda)\to\lambda} : l(\lambda < 1+2\} \cup \{w_1 \ldots w_k\} \cup \{p\}$ ($w_1,\ldots,w_k$ are the variables introduced when eliminating the combinators, including the lists $\vec{u}$ and $\vec{v}$), and $q_1 \ldots q_r$ may be composite. The type of $q\,q_1 \ldots q_r$ is 0.

We will prove by induction on the number of occurrences of p in $q\,q_1 \ldots q_r$ that $FP(\lambda p\, \lambda\vec{v}\,.\,q\,q_1 \ldots q_r)$ can be expressed as an explicit definition in which there is no $FP^{(\lambda\to\lambda)\to\lambda}$ with $l(\lambda) = 1+2$.

i. p does not occur in $q\,q_1 \ldots q_r$. Then $FP(\lambda p\, \lambda\vec{v}.\, q\,q \ldots q_r) = \lambda\vec{v}\,.\,q\,q_1 \ldots q_r$.

ii. Suppose that we have proved the result when the number of occurrences of p is $\leq k$. Suppose that there are $k+1$ occurrences of p in $q\,q_1 \ldots q_r$.

How can p occur in $q q_1 \dots q_r$? Each occurrence has the form $\lambda\vec{w} . p f_2 \dots f_n$ $(= g)$ where the lists $\vec{w}$, $f_2 \dots f_n$ may be empty. Let ν be the type of g. Can $l(\nu) \geq l+2$? If $l(\nu) \geq l+2$ then g is not $q q_1 \dots q_r$ because the type of $q q_1 \dots q_r$ is 0. Hence g occurs as $r_1 r_2 \dots r_k g r_{k+2} \dots r_n$, where r_1 is initial. The level of the type of r_1 is above the level of g, hence above $l+2$. The only initial objects available of this type are $FP^{(\lambda\to\lambda)\to\lambda}$ with $l(\lambda) = l+1$. So $r_1 = FP^{(\lambda\to\lambda)\to\lambda}$ where $l(\lambda) = l+1$. If $k \geq 2$ then g is not r_2, and the level of the type of $r_1 r_2$ is above the level of g, i.e., above $l+2$. But the type of $r_1 r_2$ is λ where $l(\lambda) = l+1$. Hence $k = 1$, i.e. $r_2 = g$. So g occurs as $FP^{(\lambda\to\lambda)\to\lambda} g$ $r_3 \dots r_n$.

Pick out an occurrence of p of the form $\lambda\vec{w} . p f_2 \dots f_n$ such that there is no occurrence of p in $f_2 \dots f_n$. Let g and ν be as above. If $l(\nu) < l+2$ then substitute g by a new variable y of type ν. If $l(\nu) \geq l+2$ then g appears as $FP^{(\lambda\to\lambda)\to\lambda} g\, r_3 \dots r_n$. Substitute $FP^{(\lambda\to\lambda)\to\lambda} g$ by a new variable y of type λ. Note that in any case the level of the type of y is strictly less that $l+2$. Let q' be the explicit definition thus obtained from $q q_1 \dots q_r$. The number of occurrences of p in q' is k. By the induction hypothesis there is an explicit definition for $FP^{(\tau\to\tau)\to\tau}(\lambda p \lambda\vec{v} . q')$ in which there is no occurrence of an $FP^{(\lambda\to\lambda)\to\lambda}$ with $l(\lambda) = l+2$. Let T be the hereditarily consistent object such that $T y$ —— $=$ $FP^{(\tau\to\tau)\to\tau}(\lambda p \lambda\vec{v} . q')$, where —— is the list of all free variables in $\lambda p \lambda\vec{v} . q'$ except y. Note that $T \in \mathcal{R}_{l+1}(\mathcal{B})$.

We will obtain an explicit definition for $FP^{(\tau\to\tau)\to\tau}(\lambda p \lambda\vec{v} . q q_1 \dots q_r)$ by substituting an object y_0 for y in $T y$ ——. This object will be in $\mathcal{R}_{l+1}(\mathcal{B})$. Suppose we have already done this substitution. Then $T y_0$ —— is the least fixed point p_0 for $\lambda p \lambda\vec{v} . q q_1 \dots q_r$. Suppose we replaced $\lambda\vec{w} . p f_2 \dots f_n$ by y. Then $y_0 = \lambda\vec{w} . p_0 f_2 \dots f_n =$ $\lambda\vec{w} . T y_0$ —— $f_2 \dots f_n$. Hence y_0 is a fixed point for

$\lambda y . \lambda \vec{w} . Ty \text{---} f_2 \dots f_n$. This can be utilized to define y_o. In this case let $y_o = FP(\lambda y \lambda \vec{w} . Ty \text{---} f_2 \dots f_n)$. In the other case we substituted $FP^{(\lambda\to\lambda)\to\lambda}(\lambda\vec{w} . p f_2 \dots f_n)$ by y, where $l(\lambda) = l+1$. Let $y_o = FP(\lambda y . FP^{(\lambda\to\lambda)\to\lambda}(\lambda\vec{w} . Ty \text{---} f_2 \dots f_n))$. Note that in each of these cases $FP = FP^{(\nu\to\nu)\to\nu}$ with $l(\nu) < l+2$. (It is possible that $FP = FP^{(o\to o)\to o}$, but this FP is in $\mathcal{R}_{l+1}(\mathcal{B})$ by proposition 2 and the observation following this proposition.) Hence $y_o \in \mathcal{R}_{l+1}(\mathcal{B})$. Let $p'_o = Ty_o \text{---}$. Then $p'_o \in \mathcal{R}_{l+1}(\mathcal{B})$. Let $p_o = FP^{(\tau\to\tau)\to\tau}(\lambda p \lambda\vec{v} . q q_1 \dots q_r)$.

It remains to prove that $p_o = p'_o$. For conviency we introduce the hereditarily consistent objects X, Y such that $Xp \text{---} = \lambda\vec{v} . q q_1 \dots q_r$, $Y p y \text{---} = \lambda\vec{v} . q'$, where the list --- is as in the definition of T. Let $p^o = \uparrow^\tau$, $p^{\alpha+1} = X p^\alpha \text{---}$, $p^\lambda = \bigcup_{\beta<\lambda} p^\beta$ if λ is a limit ordinal. Then $p_o = \bigcup\{p^\beta : \beta \in On\}$ (On is the class of ordinals). Let $(p')^o = \uparrow^\tau$, $(p')^{\alpha+1} = Y(p')^\alpha y \text{---}$, $(p')^\lambda = \bigcup_{\beta<\lambda} (p')^\beta$ if λ is a limit ordinal. Let $p' = \bigcup\{(p')^\beta : \beta \in On\}$. Then $p' = Ty \text{---}$. Let Z be a hereditarily consistent object such that $Zp \text{---}$ is the object we replaced by y (i.e. $Zp \text{---} = \lambda\vec{w} . p f_2 \dots f_n$, or $Zp \text{---} = FP^{(\lambda\to\lambda)\to\lambda}(\lambda\vec{w} . p f_2 \dots f_n))$. Then $y_o = FP(\lambda y . Z(Ty \text{---}) \text{---})$.

<u>Claim 1</u>: If $y \le Zp_o \text{---}$ then $p' \le p_o$.

Proof: We prove by induction on β that $(p')^\beta \le p_o$. This is obvious when $\beta = 0$. Suppose $\beta = \alpha + 1$, and $(p')^\alpha \le p_o$. Then $(p')^\beta = Y(p')^\alpha y \text{---} \le Y p_o(Zp_o \text{---}) \text{---} = X p_o \text{---}$ (because $Y p(Zp \text{---}) \text{---} = X p \text{---}$ for any p). But p_o is a fixed point for $\lambda p . Xp \text{---}$, hence $X p_o \text{---} = p_o$, hence $(p')^\beta \le p_o$.

<u>Claim 2</u>: $y_o \le Z p_o \text{---}$.

Proof: Let $(y_o)^o = \uparrow^\nu$ (ν is the type of y_o), $(y_o)^{\alpha+1} = Z(T(y_o)^\alpha \text{---}) \text{---}$, $(y_o)^\lambda = \bigcup_{\beta<\lambda} (y_o)^\beta$ if λ is a limit ordinal. Then $y_o = \bigcup\{(y_o)^\beta : \beta \in On\}$. We prove by induction on β that $(y_o)^\beta \le Zp_o \text{---}$.

This is obvious when $\beta = 0$. Suppose $\beta = \alpha + 1$ and $(y_o)^\alpha \leq Z p_0 ——$. Then $(y_o)^{\alpha+1} = Z(T(y_o)^\alpha ——) —— \leq Z(T(Z p_0 ——)——)——$. Now $T(Z p_0 ——) —— = p_0$ by the definition of T and Z. Hence $(y_o)^{\alpha+1} \leq Z p_0 ——$.

Claim 3: $p_0' \leq p_0$.

Proof: $p_0' = T y_o ——$. By claim 2 $y_o \leq Z p_0 ——$. By claim 1 $p_0' \leq p_0$.

Claim 4: If $Z p^\alpha —— \leq y$ then $p^{\alpha+1} \leq p'$.

Proof: We prove by induction on β that $p^\beta \leq p'$. This is obvious when $\beta = 0$. Suppose $\beta = \gamma + 1$, $\gamma \leq \alpha$ and $p^\gamma \leq p'$. Then $p^{\gamma+1} = X p^\gamma ——$. If $Z p —— \leq y$ then $X p —— \leq Y p y ——$ for any p. By assumption $Z p^\alpha —— \leq y$. Hence $Z p^\gamma —— \leq y$ because $Z p^\gamma —— \leq Z p^\alpha ——$. Hence $X p^\gamma —— \leq Y p^\gamma y ——$. By the induction hypothesis $p^\gamma \leq p'$. Hence $Y p^\gamma y —— \leq Y p' y ——$. p' is the least fixed point of $\lambda p . Y p y ——$. Hence $Y p' y —— = p'$. Hence $p^{\gamma+1} \leq p'$.

Claim 5: If $p^\alpha \leq p_0'$ then $Z p^\alpha —— \leq y_o$.

Proof: $Z p^\alpha —— \leq Z p_0' ——$. $p_0' = T y_o ——$, and $y_o = Z(T y_o ——)——$. Hence $Z p_0' —— = Z(T y_o ——) —— = y_o$. Hence $Z p^\alpha —— \leq y_o$.

Claim 6: $p_0 \leq p_0'$.

Proof: We prove by induction on β that $p^\beta \leq p_0'$. This is obvious when $\beta = 0$. Suppose $\beta = \alpha + 1$ and $p^\alpha \leq p_0'$. By claim 5 $Z p^\alpha —— \leq y_o$. By claim 4 $p^{\alpha+1} \leq p_0'$. Hence $p^\beta \leq p_0'$.

Proposition 4 follows from the claims 3 and 6. This proves theorem 29. □

The main idea in this proof is illustrated by the following example. Suppose $p_0 = FP^{(\tau\to\tau)\to\tau}(\lambda p\, \lambda \vec{v} \,.\, h(p u_1 \ldots u_k, ——))$, where $h \in \mathcal{B}$ and p does not occur in ——. Let μ be the type of h. Then $l(\mu) \leq l+2$. Let ν be the type of $pu_1 \dots u_k$. Then ν is a factor in μ, hence $l(\nu) \leq l+1$. We can use $FP^{(\nu\to\nu)\to\nu}$ to extract $pu_1 \dots u_k$. This FP has lower type than $FP^{(\tau\to\tau)\to\tau}$. We get back p from $pu_1 \dots u_k$ by $p = \lambda u_1 \dots u_k . pu_1 \dots u_k$. If the level of h had been $l+3$ we could not in general replace $FP^{(\tau\to\tau)\to\tau}$ by an FP of lower type, because $l(\nu)$ might be $l+2$.

§ 13 A FINAL COMMENT CONCERNING THE TWO TYPES

Some of the results in this paper depend on the fact that the universe of the computation domain consists of two types, some results are independent of this. Below follows a short review of the paper, where special attention is paid to this dependence.

Let $\mathfrak{J}$ be the structure $(I, N, +1, M, K, L)$ where $N \subseteq I$ is a copy of the natural numbers with successor function $+1$, M is a pairing function on I with inverse functions K and L. This structure is more general than the structure in § 1 since the subindividuals are not given a priori. Given a list $\mathcal{L}$ of relations, functions and functionals one can develop recursion theory on $\mathfrak{J}$ relative to $\mathcal{L}$, as it is done in § 2. Theorem 1 is still true. The list $\mathcal{L}$ is normal if the equality relation on I is recursive in $\mathcal{L}$, and I is weakly finite. Theorems 2 and 3 can be proved as in § 4. In the proof of theorem 4 we used the fact that countably many elements in I can be coded as one element in I in an $\mathcal{L}$-recursive way. This is true because the computation domain consists of two types. The same fact is used in the proofs of theorems 5 and 6. The type structure is also essential in the proof of theorem 7. The following facts are used: If $X \subseteq I$ is a set indexed by S then all the elements in X can be coded by one element in I in an $\mathcal{L}$-recursive way. The set of prewellorderings with domain $\subseteq S$ is recursive in $\mathcal{L}$. (The last statement is a corollary of the first.)

One can define the notion of a normal computation theory on $\mathfrak{J}$ almost as in § 6 and § 8. Lemma 20 is still true. To prove lemma 21 we essentially use the fact that the relation "x is a prewellordering with domain $\subseteq S$" is Θ-computable. Hence the type structure is needed. Lemma 22 is independent of the type structure. In the proof of lemma 23 we use the fact that S is strongly finite, which is an assumption for a normal computation theory. Hence this lemma is independent of the type structure. Lemmas 24 and 25 are also independent. In the proof of theorem 8 we use the fact that the relation "x is a prewellordering with domain $\subseteq$ S" is Θ-computable. So the type structure is needed. Lemma 21, theorem 8 and the corollary of theorem 8 are used in the proof of theorem 9. Theorem 10 is independent of the type structure, as are the results in § 9 and the first results in § 10. However the two-sorted structure is essential in the results involving countable recursion structures, because then we need to code countably many individuals by one individual. In § 11 the two types are needed to prove the existence of gaps, and also to prove the results about gaps. The results in § 12 hold for any computation domain.

References

[1] D. Cenzer, Ordinal recursion and inductive definitions, in: J.E. Fenstad and P.G. Hinman (eds.), Generalized Recursion Theory (North-Holland, Amsterdam 1974) 221 - 264.

[2] J.E. Fenstad, On axiomatizing recursion theory, in: J.E. Fenstad and P.G. Hinman (eds.), Generalized Recursion Theory (North-Holland, Amsterdam 1974) 385-404.

[3] J.E. Fenstad, Computation theories: An axiomatic approach to recursion on general structures, in: Lecture Notes in Mathematics no. 499 (Springer-Verlag 1975) 143-168.

[4] R.O. Gandy, General recursive functionals of finite type and hierarchies of functionals, in: Ann. Fac. Sci. Univ. Clermont-Ferrand No. 35 (1967) 5-24.

[5] T.J. Grilliot, Hierarchies based on objects of finite type. Jour. Symb. Log. 34 (1969) 177-182.

[6] T.J. Grilliot, Selection functions for recursive functionals. Notre Dame Jour. Formal Log. X (1969) 225-234.

[7] L.A. Harrington, Contributions to recursion theory on higher types, Ph.D. Thesis, Massachusetts Institute of Technology (1973).

[8] L.A. Harrington, A.S. Kechris, On characterizing Spector classes, Jour. Symb. Log. 40 (1975) 19-24.

[9] L.A. Harrington, A.S. Kechris, S.G. Simpson, 1-envelopes of type 2 objects, American Math. Soc. Notices 20 (1973) A-587.

[10] L.A. Harrington, D.B. MacQueen, Selection in abstract recursion theory, Journ. Symb. Log. 41 (1976) 153-158.

[11] A.S. Kechris, The structure of envelopes: A survey of recursion in higher types. M.I.T. Logic Seminar Notes, December 1973.

[12] S.C. Kleene, Recursive functionals and quantifiers of finite types I, Trans. Amer. Math. Soc. 91 (1959) 1-52; and II 108 (1963) 106-142.

[13] D.B. MacQueen, Post's problem for recursion in higher types, Ph.D. Thesis, Massachusetts Institute of Technology, 1972.

[14] Y.N. Moschovakis, Hyperanalytic predicates, Trans. Amer. Math. Soc. 129 (1967) 249-282.

[15] Y.N. Moschovakis, Abstract first order computatility I, Trans. Amer. Math. Soc. 138 (1969) 427-464, and II 138 (1969) 465-504.

[16] Y.N. Moschovakis, Axioms for computation theories - first draft, in: R.O. Gandy and C.E.M. Yates (eds.), Logic Colloquium '69 (North-Holland, Amsterdam 1971) 199-255.

[17] Y.N. Moschovakis, Structural characterizations of classes of relations, in: J.E. Fenstad and P.G. Hinman (eds.), Generalized Recursion Theory (North-Holland, Amsterdam 1974) 53-79.

[18] Y.N. Moschovakis, Elementary induction on abstract structures (North-Holland, Amsterdam 1974).

[19] D. Normann, Imbedding of higher type theories, Preprint Series No. 16 1974, Univ. of Oslo.

[20] G.E. Sacks, The 1-section of a type n object, in: J.E. Fenstad and P.G. Hinman (eds.), Genralized Recursion Theory (North-Holland, Amsterdam 1974) 81-93.

[21] G.E. Sacks, The k-section of a type n object, to appear.

[22] J.R. Shoenfield, A hierarchy based on a type two object, Trans. Amer. Math. Soc. 134 (1968) 103-108.

[23] J. Bergstra, Computability and continuity in finite types, Doctorial dissertation, University of Utrecht,1976.

[24] J. Moldestad and D. Normann, Models for recursion theory, to appear in Jour. Symb. Log.

[25] Y.N. Moschovakis, Descriptive Set Theory, to appear (North-Holland, Amsterdam).

[26] R.A. Platek, Foundations of recursion theory, Stanford University, January 1966, not published.

[27] K.J. Devlin, Aspects of constructibility, Lecture Notes in Mathematics no. 354 (Springer-Verlag 1973).

Index of notations

Subject index

Subject

Vol. 399: Functional Analysis and its Applications. Proceedings 1973. Edited by H. G. Garnir, K. R. Unni and J. H. Williamson. II, 584 pages. 1974.

Vol. 400: A Crash Course on Kleinian Groups. Proceedings 1974. Edited by L. Bers and I. Kra. VII, 130 pages. 1974.

Vol. 401: M. F. Atiyah, Elliptic Operators and Compact Groups. V, 93 pages. 1974.

Vol. 402: M. Waldschmidt, Nombres Transcendants. VIII, 277 pages. 1974.

Vol. 403: Combinatorial Mathematics. Proceedings 1972. Edited by D. A. Holton. VIII, 148 pages. 1974.

Vol. 404: Théorie du Potentiel et Analyse Harmonique. Edité par J. Faraut. V, 245 pages. 1974.

Vol. 405: K. J. Devlin and H. Johnsbråten, The Souslin Problem. VIII, 132 pages. 1974.

Vol. 406: Graphs and Combinatorics. Proceedings 1973. Edited by R. A. Bari and F. Harary. VIII, 355 pages. 1974.

Vol. 407: P. Berthelot, Cohomologie Cristalline des Schémas de Caracteristique $p > o$. II, 604 pages. 1974.

Vol. 408: J. Wermer, Potential Theory. VIII, 146 pages. 1974.

Vol. 409: Fonctions de Plusieurs Variables Complexes, Séminaire François Norguet 1970–1973. XIII, 612 pages. 1974.

Vol. 410: Séminaire Pierre Lelong (Analyse) Année 1972–1973. VI, 181 pages. 1974.

Vol. 411: Hypergraph Seminar. Ohio State University, 1972. Edited by C. Berge and D. Ray-Chaudhuri. IX, 287 pages. 1974.

Vol. 412: Classification of Algebraic Varieties and Compact Complex Manifolds. Proceedings 1974. Edited by H. Popp. V, 333 pages. 1974.

Vol. 413: M. Bruneau, Variation Totale d'une Fonction. XIV, 332 pages. 1974.

Vol. 414: T. Kambayashi, M. Miyanishi and M. Takeuchi, Unipotent Algebraic Groups. VI, 165 pages. 1974.

Vol. 415: Ordinary and Partial Differential Equations. Proceedings 1974. XVII, 447 pages. 1974.

Vol. 416: M. E. Taylor, Pseudo Differential Operators. IV, 155 pages. 1974.

Vol. 417: H. H. Keller, Differential Calculus in Locally Convex Spaces. XVI, 131 pages. 1974.

Vol. 418: Localization in Group Theory and Homotopy Theory and Related Topics. Battelle Seattle 1974 Seminar. Edited by P. J. Hilton. VI, 172 pages 1974.

Vol. 419: Topics in Analysis. Proceedings 1970. Edited by O. E. Lehto, I. S. Louhivaara, and R. H. Nevanlinna. XIII, 392 pages. 1974.

Vol. 420: Category Seminar. Proceedings 1972/73. Edited by G. M. Kelly. VI, 375 pages. 1974.

Vol. 421: V. Poénaru, Groupes Discrets. VI, 216 pages. 1974.

Vol. 422: J.-M. Lemaire, Algèbres Connexes et Homologie des Espaces de Lacets. XIV, 133 pages. 1974.

Vol. 423: S. S. Abhyankar and A. M. Sathaye, Geometric Theory of Algebraic Space Curves. XIV, 302 pages. 1974.

Vol. 424: L. Weiss and J. Wolfowitz, Maximum Probability Estimators and Related Topics. V, 106 pages. 1974.

Vol. 425: P. R. Chernoff and J. E. Marsden, Properties of Infinite Dimensional Hamiltonian Systems. IV, 160 pages. 1974.

Vol. 426: M. L. Silverstein, Symmetric Markov Processes. X, 287 pages. 1974.

Vol. 427: H. Omori, Infinite Dimensional Lie Transformation Groups. XII, 149 pages. 1974.

Vol. 428: Algebraic and Geometrical Methods in Topology, Proceedings 1973. Edited by L. F. McAuley. XI, 280 pages. 1974.

Vol. 429: L. Cohn, Analytic Theory of the Harish-Chandra C-Function. III, 154 pages. 1974.

Vol. 430: Constructive and Computational Methods for Differential and Integral Equations. Proceedings 1974. Edited by D. L. Colton and R. P. Gilbert. VII, 476 pages. 1974.

Vol. 431: Séminaire Bourbaki – vol. 1973/74. Exposés 436–452. IV, 347 pages. 1975.

Vol. 432: R. P. Pflug, Holomorphiegebiete, pseudokonvexe Gebiete und das Levi-Problem. VI, 210 Seiten. 1975.

Vol. 433: W. G. Faris, Self-Adjoint Operators. VII, 115 pages. 1975.

Vol. 434: P. Brenner, V. Thomée, and L. B. Wahlbin, Besov Spaces and Applications to Difference Methods for Initial Value Problems. II, 154 pages. 1975.

Vol. 435: C. F. Dunkl and D. E. Ramirez, Representations of Commutative Semitopological Semigroups. VI, 181 pages. 1975.

Vol. 436: L. Auslander and R. Tolimieri, Abelian Harmonic Analysis, Theta Functions and Function Algebras on a Nilmanifold. V, 99 pages. 1975.

Vol. 437: D. W. Masser, Elliptic Functions and Transcendence. XIV, 143 pages. 1975.

Vol. 438: Geometric Topology. Proceedings 1974. Edited by L. C. Glaser and T. B. Rushing. X, 459 pages. 1975.

Vol. 439: K. Ueno, Classification Theory of Algebraic Varieties and Compact Complex Spaces. XIX, 278 pages. 1975

Vol. 440: R. K. Getoor, Markov Processes: Ray Processes and Right Processes. V, 118 pages. 1975.

Vol. 441: N. Jacobson, PI-Algebras. An Introduction. V, 115 pages. 1975.

Vol. 442: C. H. Wilcox, Scattering Theory for the d'Alembert Equation in Exterior Domains. III, 184 pages. 1975.

Vol. 443: M. Lazard, Commutative Formal Groups. II, 236 pages. 1975.

Vol. 444: F. van Oystaeyen, Prime Spectra in Non-Commutative Algebra. V, 128 pages. 1975.

Vol. 445: Model Theory and Topoi. Edited by F. W. Lawvere, C. Maurer, and G. C. Wraith. III, 354 pages. 1975.

Vol. 446: Partial Differential Equations and Related Topics. Proceedings 1974. Edited by J. A. Goldstein. IV, 389 pages. 1975.

Vol. 447: S. Toledo, Tableau Systems for First Order Number Theory and Certain Higher Order Theories. III, 339 pages. 1975.

Vol. 448: Spectral Theory and Differential Equations. Proceedings 1974. Edited by W. N. Everitt. XII, 321 pages. 1975.

Vol. 449: Hyperfunctions and Theoretical Physics. Proceedings 1973. Edited by F. Pham. IV, 218 pages. 1975.

Vol. 450: Algebra and Logic. Proceedings 1974. Edited by J. N. Crossley. VIII, 307 pages. 1975.

Vol. 451: Probabilistic Methods in Differential Equations. Proceedings 1974. Edited by M. A. Pinsky. VII, 162 pages. 1975.

Vol. 452: Combinatorial Mathematics III. Proceedings 1974. Edited by Anne Penfold Street and W. D. Wallis. IX, 233 pages. 1975.

Vol. 453: Logic Colloquium. Symposium on Logic Held at Boston, 1972–73. Edited by R. Parikh. IV, 251 pages. 1975.

Vol. 454: J. Hirschfeld and W. H. Wheeler, Forcing, Arithmetic, Division Rings. VII, 266 pages. 1975.

Vol. 455: H. Kraft, Kommutative algebraische Gruppen und Ringe. III, 163 Seiten. 1975.

Vol. 456: R. M. Fossum, P. A. Griffith, and I. Reiten, Trivial Extensions of Abelian Categories. Homological Algebra of Trivial Extensions of Abelian Categories with Applications to Ring Theory. XI, 122 pages. 1975.

Vol. 457: Fractional Calculus and Its Applications. Proceedings 1974. Edited by B. Ross. VI, 381 pages. 1975.

Vol. 458: P. Walters, Ergodic Theory – Introductory Lectures. VI, 198 pages. 1975.

Vol. 459: Fourier Integral Operators and Partial Differential Equations. Proceedings 1974. Edited by J. Chazarain. VI, 372 pages. 1975.

Vol. 460: O. Loos, Jordan Pairs. XVI, 218 pages. 1975.

Vol. 461: Computational Mechanics. Proceedings 1974. Edited by J. T. Oden. VII, 328 pages. 1975.

Vol. 462: P. Gérardin, Construction de Séries Discrètes p-adiques. »Sur les séries discrètes non ramifiées des groupes réductifs déployés p-adiques«. III, 180 pages. 1975.

Vol. 463: H.-H. Kuo, Gaussian Measures in Banach Spaces. VI, 224 pages. 1975.

Vol. 464: C. Rockland, Hypoellipticity and Eigenvalue Asymptotics. III, 171 pages. 1975.

Vol. 465: Séminaire de Probabilités IX. Proceedings 1973/74. Edité par P. A. Meyer. IV, 589 pages. 1975.

Vol. 466: Non-Commutative Harmonic Analysis. Proceedings 1974. Edited by J. Carmona, J. Dixmier and M. Vergne. VI, 231 pages. 1975.

Vol. 467: M. R. Essén, The Cos $\pi\lambda$ Theorem. With a paper by Christer Borell. VII, 112 pages. 1975.

Vol. 468: Dynamical Systems – Warwick 1974. Proceedings 1973/74. Edited by A. Manning. X, 405 pages. 1975.

Vol. 469: E. Binz, Continuous Convergence on C(X). IX, 140 pages. 1975.

Vol. 470: R. Bowen, Equilibrium States and the Ergodic Theory of Anosov Diffeomorphisms. III, 108 pages. 1975.

Vol. 471: R. S. Hamilton, Harmonic Maps of Manifolds with Boundary. III, 168 pages. 1975.

Vol. 472: Probability-Winter School. Proceedings 1975. Edited by Z. Ciesielski, K. Urbanik, and W. A. Woyczyński. VI, 283 pages. 1975.

Vol. 473: D. Burghelea, R. Lashof, and M. Rothenberg, Groups of Automorphisms of Manifolds. (with an appendix by E. Pedersen) VII, 156 pages. 1975.

Vol. 474: Séminaire Pierre Lelong (Analyse) Année 1973/74. Edité par P. Lelong. VI, 182 pages. 1975.

Vol. 475: Répartition Modulo 1. Actes du Colloque de Marseille-Luminy, 4 au 7 Juin 1974. Edité par G. Rauzy. V, 258 pages. 1975. 1975.

Vol. 476: Modular Functions of One Variable IV. Proceedings 1972. Edited by B. J. Birch and W. Kuyk. V, 151 pages. 1975.

Vol. 477: Optimization and Optimal Control. Proceedings 1974. Edited by R. Bulirsch, W. Oettli, and J. Stoer. VII, 294 pages. 1975.

Vol. 478: G. Schober, Univalent Functions – Selected Topics. V, 200 pages. 1975.

Vol. 479: S. D. Fisher and J. W. Jerome, Minimum Norm Extremals in Function Spaces. With Applications to Classical and Modern Analysis. VIII, 209 pages. 1975.

Vol. 480: X. M. Fernique, J. P. Conze et J. Gani, Ecole d'Eté de Probabilités de Saint-Flour IV-1974. Edité par P.-L. Hennequin. XI, 293 pages. 1975.

Vol. 481: M. de Guzmán, Differentiation of Integrals in R^n. XII, 226 pages. 1975.

Vol. 482: Fonctions de Plusieurs Variables Complexes II. Séminaire François Norguet 1974–1975. IX, 367 pages. 1975.

Vol. 483: R. D. M. Accola, Riemann Surfaces, Theta Functions, and Abelian Automorphisms Groups. III, 105 pages. 1975.

Vol. 484: Differential Topology and Geometry. Proceedings 1974. Edited by G. P. Joubert, R. P. Moussu, and R. H. Roussarie. IX, 287 pages. 1975.

Vol. 485: J. Diestel, Geometry of Banach Spaces – Selected Topics. XI, 282 pages. 1975.

Vol. 486: S. Stratila and D. Voiculescu, Representations of AF-Algebras and of the Group U (∞). IX, 169 pages. 1975.

Vol. 487: H. M. Reimann und T. Rychener, Funktionen beschränkter mittlerer Oszillation. VI, 141 Seiten. 1975.

Vol. 488: Representations of Algebras, Ottawa 1974. Proceedings 1974. Edited by V. Dlab and P. Gabriel. XII, 378 pages. 1975.

Vol. 489: J. Bair and R. Fourneau, Etude Géométrique des Espaces Vectoriels. Une Introduction. VII, 185 pages. 1975.

Vol. 490: The Geometry of Metric and Linear Spaces. Proceedings 1974. Edited by L. M. Kelly. X, 244 pages. 1975.

Vol. 491: K. A. Broughan, Invariants for Real-Generated Uniform Topological and Algebraic Categories. X, 197 pages. 1975.

Vol. 492: Infinitary Logic: In Memoriam Carol Karp. Edited by D. W. Kueker. VI, 206 pages. 1975.

Vol. 493: F. W. Kamber and P. Tondeur, Foliated Bundles and Characteristic Classes. XIII, 208 pages. 1975.

Vol. 494: A Cornea and G. Licea. Order and Potential Resolvent Families of Kernels. IV, 154 pages. 1975.

Vol. 495: A. Kerber, Representations of Permutation Groups II. V, 175 pages. 1975.

Vol. 496: L. H. Hodgkin and V. P. Snaith, Topics in K-Theory. Two Independent Contributions. III, 294 pages. 1975.

Vol. 497: Analyse Harmonique sur les Groupes de Lie. Proceedings 1973–75. Edité par P. Eymard et al. VI, 710 pages. 1975.

Vol. 498: Model Theory and Algebra. A Memorial Tribute to Abraham Robinson. Edited by D. H. Saracino and V. B. Weispfenning. X, 463 pages. 1975.

Vol. 499: Logic Conference, Kiel 1974. Proceedings. Edited by G. H. Müller, A. Oberschelp, and K. Potthoff. V, 651 pages 1975.

Vol. 500: Proof Theory Symposion, Kiel 1974. Proceedings. Edited by J. Diller and G. H. Müller. VIII, 383 pages. 1975.

Vol. 501: Spline Functions, Karlsruhe 1975. Proceedings. Edited by K. Böhmer, G. Meinardus, and W. Schempp. VI, 421 pages. 1976.

Vol. 502: János Galambos, Representations of Real Numbers by Infinite Series. VI, 146 pages. 1976.

Vol. 503: Applications of Methods of Functional Analysis to Problems in Mechanics. Proceedings 1975. Edited by P. Germain and B. Nayroles. XIX, 531 pages. 1976.

Vol. 504: S. Lang and H. F. Trotter, Frobenius Distributions in GL_2-Extensions. III, 274 pages. 1976.

Vol. 505: Advances in Complex Function Theory. Proceedings 1973/74. Edited by W. E. Kirwan and L. Zalcman. VIII, 203 pages. 1976.

Vol. 506: Numerical Analysis, Dundee 1975. Proceedings. Edited by G. A. Watson. X, 201 pages. 1976.

Vol. 507: M. C. Reed, Abstract Non-Linear Wave Equations. VI, 128 pages. 1976.

Vol. 508: E. Seneta, Regularly Varying Functions. V, 112 pages. 1976.

Vol. 509: D. E. Blair, Contact Manifolds in Riemannian Geometry. VI, 146 pages. 1976.

Vol. 510: V. Poènaru, Singularités C^∞ en Présence de Symétrie. V, 174 pages. 1976.

Vol. 511: Séminaire de Probabilités X. Proceedings 1974/75. Edité par P. A. Meyer. VI, 593 pages. 1976.

Vol. 512: Spaces of Analytic Functions, Kristiansand, Norway 1975. Proceedings. Edited by O. B. Bekken, B. K. Øksendal, and A. Stray. VIII, 204 pages. 1976.

Vol. 513: R. B. Warfield, Jr. Nilpotent Groups. VIII, 115 pages. 1976.

Vol. 514: Séminaire Bourbaki vol. 1974/75. Exposés 453 – 470. IV, 276 pages. 1976.

Vol. 515: Bäcklund Transformations. Nashville, Tennessee 1974. Proceedings. Edited by R. M. Miura. VIII, 295 pages. 1976.